Better Children's Sleep

The Parent's Guide to Better Nights: Baby Sleep, Toddler Sleep, and Beyond Backed by Pediatric Neuroscience

The Sleep Engineer, Volume 2

C. Louis-Charles, PhD

Cybersoft Publishing LLC

Cybersoft Publishing LLC

Fort Washington, MD 20744

Drclaude.net

First Edition, June 2026

A Note on Infant Safe Sleep

This book discusses sleep for children from birth through adolescence. For infants under twelve months, please follow the American Academy of Pediatrics safe-sleep guidance at all times, no matter what method, schedule, or routine you choose from this book.

The AAP recommends placing your baby on their back for every sleep on a firm, flat, level sleep surface in a crib, bassinet, or play yard that meets current safety standards. Use only a tight-fitting sheet. Keep soft objects, loose bedding, pillows, bumpers, and toys out of the sleep area. Room-share with your baby for at least the first six months, ideally up to twelve months, but do not share a sleep surface. Offer a pacifier at sleep times after breastfeeding is well established. Avoid smoke exposure, alcohol, and illicit drugs during pregnancy and after birth.

Swaddling is a useful soothing tool but must be stopped as soon as your baby shows any sign of rolling over, typically between two and four months. Once rolling begins, transition to a sleep sack or a wearable blanket with the arms free.

Nothing in this book overrides medical advice from your pediatrician. If you are unsure about any sleep setup, practice, or symptom, talk to your child's doctor before making changes. For the most current guidance, search the AAP's safe-sleep recommendations directly.

A Note on Medical Advice

This book is an educational guide to the brain science of childhood and adolescent sleep. It is not a medical document. It is not a substitute for evaluation by a pediatrician, pediatric sleep specialist, or other licensed healthcare professional. Reading it will help you understand how your child's sleep develops and how to support better nights as a family. It will not diagnose or treat any pediatric sleep disorder.

If your child snores loudly with gasping pauses, mouth-breathes through the night, falls asleep involuntarily during the day, has had severe insomnia for more than three months, or experiences any sleep symptom that worries you, please consult your pediatrician or a pediatric sleep specialist. Pediatric obstructive sleep apnea, restless legs syndrome, narcolepsy, and other conditions require formal evaluation and treatment that goes well beyond anything a book can offer.

The author and publisher disclaim any responsibility for adverse outcomes resulting from the use or misuse of information contained in this book. The reader assumes full responsibility for any actions taken based on the material presented here.

Contents

2 INTRODUCTION 1

Why most sleep books fall short 2
Who this book is for 3
The Sleep Engineer approach 5
How to read this book 6
A promise 7

3 THE TWO-BRAIN PROBLEM 10

Chapter Summary 11
What This Chapter Will Cover 12
The Body Clock You Never Knew You Were Sharing 13
When One Person's Bad Night Belongs to Everyone 17
The Cortisol-Oxytocin Loop at 3 A.M. 20
Sleep Training Is Nervous-System Training 24
Two Brains, One Rhythm 29
Try This Tonight 32
Parent Takeaways 33

4 THE NEWBORN BRAIN: FIRST 3 MONTHS 35

Chapter Summary 36
What This Chapter Will Cover 37
The Clock That Has Not Started Yet 38
Active Sleep, Quiet Sleep, and the Developing Brain 41
The 50-Minute Newborn Sleep Cycle 43
Why Sleeping Through the Night Is Not Possible Yet 45
The Fourth Trimester 47
Safe Sleep: What It Is and Why It Matters 49
Swaddling: Comfort With a Cutoff 51
Feeding, Hunger, and the Sleep Connection 53
The "Good Sleeper" Who Gets Harder 55
Protecting Your Own Sleep 56
Try This Tonight 57
Parent Takeaways 58

5 THE CLOCK STARTS TICKING 60

Chapter Summary 61
What This Chapter Will Cover 62
The New Clock and What It Needs 65

The Four-Month Progression 67
Drowsy-But-Awake and the Formation of Self-Soothing 70
Sleep Training Methods: A Clear Map 72
The Right Method Is the One You Will Use 76
Night Feeding in Months Four Through Six 77
How to Know Your Baby Is Ready 79
Common Pitfalls in This Window 80
Try This Tonight 82
Parent Takeaways 83

6 NAPS, SCHEDULES, AND THE FIRST YEAR OF REAL ROUTINE 85

Chapter Summary 86
What This Chapter Will Cover 86
The Biology of Daytime Sleep 87
Wake Windows by Age 90
Why Daytime Sleep Is Not Optional 92
The 3-to-2 Nap Transition 94
The 2-to-1 Nap Transition 96
The 1-to-0 Nap Transition 98
Schedule, Wake Window, and Cue-Based Approaches 100
Catnaps, False Starts, and the 45-Minute Intruder 103
Building a Flexible Daily Rhythm 105
The Overtired Loop and How to Break It 108
Try This Tonight 109
Parent Takeaways 110

7 THE TODDLER BEDTIME BATTLE: YEARS 1 TO 3 111

Chapter Summary 112
What This Chapter Will Cover 113
Why Toddlers Fight Sleep: The Developing Brain 114
How the Resistance Loop Forms 117
Cortisol Overshoot and the Second Wind 118
Transitional Objects: The Neuroscience of Comfort 119
Bedtime Routines That Actually Work 122
The One More Book Trap 124
The Crib-to-Bed Transition 126
Nightmares vs Night Wakings vs Stalling 128
Why Screens After Dinner Wreck Toddler Sleep 130
Holding Loving Limits 133

Try This Tonight 134
Parent Takeaways 135

8 PRESCHOOL SLEEP: AGES THREE TO FIVE 136

Chapter Summary 137
What This Chapter Will Cover 138
Sleep Architecture at Ages Three to Five 139
The Nap-Drop Window 141
Quiet Time as a Bridge 143
Nightmares: What They Are and How to Help 145
Night Terrors: A Different Animal Entirely 147
Sleepwalking and Confusional Arousals 149
Imagination, Monsters, and Bedtime Fears 152
The Monster Spray Method and Other Strategies 154
Starting Preschool and Schedule Disruptions 156
Try This Tonight 159
Parent Takeaways 160

9 SCHOOL-AGED SLEEP: AGES 6 TO 11 161

Chapter Summary 162
What This Chapter Will Cover 163
How Much Sleep Does Your Child Actually Need? 164
The Hyperactive-Not-Sleepy Paradox 166
Sleep Deprivation vs ADHD: What to Know 168
Slow-Wave Sleep and Memory Consolidation 170
The Homework-Screen-Bedtime Trap 172
Over-Scheduling and Sleep Debt 174
Social Jet Lag in School-Age Kids 176
The School-Night Routine That Works 178
Sleep and Emotional Regulation 180
Try This Tonight 183
Parent Takeaways 184

10 THE TEENAGE BRAIN 186

Chapter Summary 187
What This Chapter Will Cover 188
Why the Clock Shifts at Puberty 189
Eight to Ten Hours Is Not a Suggestion 191
The 7 a.m. Bell and What It Actually Does 193
What the AAP and AASM Have Said 195

Light as a Tool, Not Just a Problem 195
The Phone in the Bedroom 197
Caffeine and the Teen Brain 199
Social Jet Lag and the Weekend Spiral 201
Sleep Debt and Mental Health in Teens 202
Navigating This with Your Teenager 205
What You Can Do at the School Level 207
The Long View 208
Try This Tonight 209
Parent Takeaways 210

11 ROUTINES THAT HOLD 212

Chapter Summary 213
What This Chapter Will Cover 213
Why Generic Schedules Fail Real Families 214
Staggered vs Synced Bedtimes 216
The Three-or-More-Kid Problem 218
The Single-Parent Reality 220
Shift-Work Families 222
Co-Parenting Across Two Homes 224
Travel, Holidays, and Grandparent Visits 226
Bedsharing and Cosleeping 228
Recovery After a Disrupted Week 230
Building Routines That Survive Real Life 232
Try This Tonight 234
Parent Takeaways 235
What About Screens at Bedtime 236

12 ANXIETY, WORRIES, AND THE BEDTIME MIND 238

Chapter Summary 239
What This Chapter Will Cover 240
Why Bedtime Activates the Anxious Brain 241
The Default Mode Network at Night 243
Developmental Fears by Age 245
CBT-I Principles Adapted for Children 247
The Worry-Time Technique 250
Belly Breathing and the Parasympathetic Reset 252
Visualization Scripts for Children 254
When Anxiety Crosses a Clinical Line 256
Co-Regulating with an Anxious Child 258

Building a Worry-Ready Home 261
Try This Tonight 262
Parent Takeaways 263

13 TROUBLESHOOTING AND RED FLAGS 265

Chapter Summary 266
What This Chapter Will Cover 266
Pediatric Obstructive Sleep Apnea 268
Restless Legs Syndrome in Children 272
Pediatric Narcolepsy 274
Parasomnias: A Map of the Night 277
Bedwetting: When to Wait, When to Act 280
Bruxism: Teeth Grinding in Children 282
Sleep-Related Anxiety vs. Sleep-Related Medical Issues 282
When to Consult a Pediatric Sleep Specialist 284
What a Pediatric Sleep Study Involves 286
Melatonin in Children: Use With Caution 288
Try This Tonight 290
Parent Takeaways 291

14 CHILDREN'S SLEEP MYTHS DECODED 293

Chapter Summary 294
What This Chapter Will Cover 295
Myth One: Good Babies Sleep Through the Night 297
Myth Two: You Should Never Let Them Cry 299
Myth Three: Crying It Out Causes Lasting Trauma 301
Myth Four: Bedsharing Is Always Dangerous 302
Myth Five: Skipping Naps Means Better Night Sleep 304
Myth Six: TV Helps Kids Wind Down 305
Myth Seven: Teens Are Just Lazy Sleepers 306
Myth Eight: Melatonin Is Harmless for Kids 308
Myth Nine: Sleep Training Breaks the Parent-Child Bond 310
Myth Ten: My Kid Will Grow Out of It 312
Myth Eleven: More Sleep Is Always Better 313
Myth Twelve: The Perfect Schedule Exists 315
Try This Tonight 316
Parent Takeaways 317
This Is Where You Are Now 318

15 A FINAL NOTE 320

You were never broken. 321
There is no perfect parent, and no perfect sleeper 322
You have permission...323
What comes next in this series...325
For the parent reading this at 11 pm, one eye open 325

16 GLOSSARY...327

17 THE TWELVE-NIGHT FAMILY SLEEP RESET 339

Introduction

It is three in the morning. The house is dark except for the soft glow of whatever screen you are reading this on. The baby has been up for forty minutes, or maybe ninety; you have lost count. Your arms are tired in a specific way that no one warns you about before you have children, a low throb that starts at the shoulders and settles somewhere deep in your chest. You have already fed, rocked, shushed, patted, put down, picked back up, and checked the room temperature twice. Nothing worked, or something worked briefly, and now you are here again.

Or maybe your child is not a baby. Maybe you have a two-year-old who climbs out of the crib before you make it to the door, who appears at your bedside at midnight with a stuffed animal and a look of complete sincerity. Maybe you have a seven-year-old who cannot settle because something at school has taken up residence in their mind and refuses to leave quietly. Maybe you have a teenager who swears, with genuine conviction, that they are not tired at eleven o'clock, not tired at midnight, barely tired at one, and then cannot get out of bed before ten if left alone. Every one of these situations is real and exhausting in its own specific way.

You are not doing this wrong. That is the first thing this book wants to say to you, before anything else. The chaos of children's sleep is not a character flaw in you or in your child. It is a feature of human brain development. Newborns do not have a working circadian clock. Toddlers have competing developmental drives that push against every routine you build. School-aged children carry anxiety to bed the way they carry backpacks everywhere. Teenagers experience a genuine, biology-driven shift in their sleep timing that makes asking them to be in bed at nine feel, from their nervous system's point of view, as strange as asking you to fall asleep at four in the afternoon. None of that is misbehavior. All of it is neuroscience.

You deserve to know that neuroscience, not in a textbook sense, not in a way that requires a biology degree or an hour of concentration, but in the plain-English way that actually helps at three in the morning. That is what this book is for.

You can put it down now if you need to. It will be here.

Why most sleep books fall short

Walk into any bookstore and the children's sleep section will give you a problem before it gives you a solution. Half the books are built around a single method, and each one argues, implicitly or explicitly, that the other methods are either harmful or useless. One book tells you that responding to your baby every time they cry is the only compassionate path. Another tells you that responding every time is the very thing creating the problem. One tells you to be firm; another tells you that firmness is damaging. You buy one, you try it, it does not quite work, and you wonder whether you are the failure or the method is.

The other problem is clinical distance. A lot of the better-researched books are written for a reader who wants to understand the mechanism, and the mechanism matters, but the prose often lands at a remove from the reality of the thing. The reality is that you are making decisions while sleep-deprived, in the middle of the night, while a small human who cannot yet explain their needs is communicating those needs as loudly as possible. A chapter that reads well at your kitchen table in daylight needs to still make sense at two in the morning, when your prefrontal cortex has had about as much rest as your child's.

The third problem is one-size reasoning applied to enormous diversity. Children are not the same. Temperament varies enormously, and it is real. Some infants fuss for thirty seconds and settle. Others need a great deal more help with that transition. Some toddlers tolerate changes in routine well; others unravel completely. Some families have the flexibility to respond consistently and at length; others are running on shift schedules, single-parent logistics, or childcare constraints that make the idealized bedtime routine in chapter one of the idealized sleep book genuinely impossible. A book that does not acknowledge that will make many parents feel like they are failing a system that was never designed for them.

There is also the problem of pace. Most sleep books are written for readers who have time to sit with them. They build slowly toward a recommendation, often burying the most actionable guidance in a later chapter while the earlier chapters establish ideological ground. That structure works fine on a Sunday afternoon; it does not work well for the parent holding the book open with one hand at midnight, because the situation is urgent and they need to know what to try tonight. This book is organized to reward both kinds of reading: the cover-to-cover reader who wants the whole framework, and the reader who needs the relevant section right now.

This book was written as a response to all three of those problems. It does not pick a side in the sleep-training debate. It presents the actual evidence on multiple approaches, explains what each method asks of your child's nervous system and what it asks of yours, and then lets you decide. It is written in prose that tries to stay readable when you are running on four hours of sleep. And it is built around the understanding that your household is specific. You will find the science here; you will find the options; the choices are yours.

Who this book is for

This book is for parents and caregivers of children from birth through age 18. That is a wide range, and deliberately so. Most sleep books pick a lane: the baby-sleep book, the toddler-sleep book. The problem with lane-picking is that families do not stay in a lane. The parent reading the newborn chapter tonight will be the parent of a two-year-old in twenty-four months. Children move through the developmental windows in a sequence, and the decisions you make in one window shape the sleep patterns that show up in the next one. Seeing the whole arc matters.

It is for couples who are disagreeing about sleep training and need a calm, evidence-based look at what the research actually shows, not what a particular parenting philosophy insists the research shows. It is for the parent who does bedtime alone because the other parent works nights, works abroad, or is no longer in the picture. It is for grandparents who are raising grandchildren and feel caught between what they were told as parents and what everyone around them seems to be doing now. It is for blended families, for households where two children of different ages need different things at the same bedtime, for families where one child sleeps easily. Another has never slept easily, and the contrast is its own kind of difficult.

It is for the parent who has already read three sleep books and feels more confused than before they started reading them. It is for the parent who has not read any sleep books because every one they picked up felt as if it were written for a family with more time, more support, or more certainty than they have.

This book is also for parents navigating different philosophies within their own households. It is common for two caregivers to arrive at the sleep problem from different directions: one convinced that responsiveness is paramount, the other convinced that limits and consistency matter more. Both of those convictions have real support in the research, and they are not as incompatible as the louder voices in the parenting world suggest. You will find here a framework that makes space for both values, and a way of talking about the evidence that does not require either parent to feel defeated.

No neuroscience background is needed here. Where scientific terms are necessary, they are defined immediately in plain language. Where the research is messy or contested, that is stated clearly rather than smoothed over. This book does not ask you to pretend the evidence is simpler than it is; it asks you to trust that you can handle the real complexity, and then it tries to make that complexity as navigable as possible.

The Sleep Engineer approach

The Sleep Engineer series rests on the idea that sleep is a system, and systems can be understood and improved. That idea applies to adult sleep, which is what Book 1 of this series covers, and it applies to children's sleep, with the important addition that children's systems are still being built. The developing brain is not a small adult brain. It has different architecture, timing, vulnerabilities, and responsiveness to the same interventions. Understanding the development is the core of the approach.

This book leans gently and responsively. That is a choice made with intent, and it deserves a brief explanation. The neuroscience of early childhood does not indicate that any one sleep-training method is universally correct. What it does tell us is that very young children regulate their nervous systems in partnership with a caregiver, a process called co-regulation, and that approaches which preserve that relational quality tend to produce outcomes that parents and children tolerate well over time. That is the lean. It is not a rule; it is a direction of emphasis.

Extinction-based methods, commonly called cry-it-out and graduated extinction, are presented here with the same intellectual honesty. The evidence for their effectiveness is real. The research on long-term harm is far weaker than either critics or proponents tend to represent. If extinction methods are a good fit for your family's situation and values, you will find the science behind them here without judgment, and without the quiet condescension that some gentler-leaning books insert when they get to that section.

The organizing principle is this: the best sleep method for your child is the one you can actually implement consistently over multiple nights, in your real household, with the support you actually have. Consistency matters more than methodology. A gentle method applied erratically is harder on a child's nervous system than a firmer method applied calmly and predictably. This book will help you understand the mechanisms; it will not tell you which mechanism is morally correct.

Evidence-based also means being honest about limits. Sleep science is a young field, especially pediatric sleep science, which has historically received far less research attention than adult sleep. Some findings are robust and have been replicated across multiple studies and populations. Others rest on smaller samples, or on populations that skew heavily toward one demographic and may not represent the full range of families. Where the evidence is strong, this book makes that clear. Where the evidence is preliminary, or where expert opinion and popular parenting culture have outrun what the data actually show, that is stated as well. You are an adult making decisions for your child, in your specific circumstances; you deserve the real picture, not a curated one that has been tidied up to avoid making you anxious.

How to read this book

The twelve chapters are organized by developmental window, moving from newborn through teenager. Chapter 1 covers the architecture of children's sleep at the level of the brain: the circadian system, sleep pressure, sleep stages, and how all of these change from birth to adolescence. It is the foundation for everything that follows. If you are the kind of reader who wants context before detail, start at Chapter 1 and go forward. Everything in the later chapters will be more legible if you have Chapter 1 behind you.

If you are in the middle of a specific, urgent situation, you can go directly to the chapter for your child's current age band. The chapters are written to stand alone. You will not be lost if you jump to the toddler chapter or the school-age chapter without reading everything before it. Each chapter reintroduces the core concepts it builds on, so you get the context you need even if you came in sideways.

If you have multiple children at different ages, the best path is to read the book in order. Watching the developmental progression from newborn to teenager as a continuous story makes it much easier to understand why a strategy that worked at six months stops working at fourteen months, and why the approach that helps a seven-year-old would be completely wrong for a two-year-old. The sequence is not arbitrary.

At the end of the book are three chapters that operate at the household level rather than the age-specific level. Chapter 9 covers routines for real families, including multi-child households, shift-work schedules, and the logistical complications that make idealized bedtime choreography difficult. Chapter 10 covers anxiety and the bedtime mind, which is relevant at every age but becomes especially important from school age onward. Chapter 11 is troubleshooting and red flags, the chapter you want if you have tried the reasonable things and something is still not working, or if you want to know when a sleep problem crosses into territory that warrants a pediatrician's involvement.

Chapter 12 closes the book with a myth decoder, a systematic look at the most persistent and confident claims in children's sleep advice, examined against the evidence. The goal is to leave you with the tools to evaluate the next thing you read or hear, so that this book is not just twelve chapters you accepted on authority but a way of thinking you can use going forward.

Each chapter includes a brief opening scenario, a summary of its contents, a practical section called Try This Tonight, and a set of parent takeaways. The structure is consistent so that when you need to come back to a chapter later, the information is where you expect it to be.

A promise

There is a particular kind of exhaustion that comes from feeling judged while you feel depleted. Parents of young children know it well. It lives in the comments from relatives who believe whatever worked in 1987 is the correct and only approach. It lives in the mommy-group thread where someone's child sleeps twelve hours, and yours has not slept twelve consecutive hours since November. It lives in the books that open with a philosophy and then make every parent who chose differently feel like a cautionary tale.

This book is not that. You will not be shamed here for how you fed your baby, for whether you co-slept, for whether you sleep-trained or did not sleep-train or started sleep training and stopped. You will not be asked to believe that your child's future cognitive development hinges on the specific method you chose at four months. The research does not actually support that level of alarm, and this book will not pretend otherwise.

What you will find here is the science of how your child's brain sleeps at each age, explained clearly and honestly. You will find the options for working with that science in your specific household, presented without a hidden agenda about which option is correct. You will find the flags that tell you when something is worth bringing to a professional, and the reassurance that many things that feel alarming are developmentally normal. You will find a perspective that takes your sleep seriously alongside your child's sleep, because a parent running on nothing cannot run a steady bedtime for anyone else. Parent sleep is not a footnote in this book; it is a variable in the system, one that affects everything from the consistency of your child's routine to your own capacity to stay calm during the hard stretches of a new method.

You do not have to be a perfect sleeper to help your child sleep better. You do not need a perfect routine, a perfect bedroom environment, or perfect consistency. You have to understand the system well enough to make reasonable decisions and to correct course when the first approach does not work, because some adjustment is almost always needed. Sleep is not a linear problem with a single correct solution applied once; it is an ongoing conversation between your child's developing nervous system and the environment you provide. What you learn about your child at four months will not cover everything you need at fourteen months. The science in this book gives you a way to keep reading that conversation accurately as the child in front of you continues to change.

You are not failing. You are learning a system that nobody trained you for, using a brain that is also sleep-deprived. That is harder than it sounds, and you are doing it anyway. Let's get into the science.

1 The Two-Brain Problem

It is 3:12 in the morning, and you are standing in the hallway outside your child's room in the dark, one hand braced against the wall, trying to remember whether you have already gone in or are about to. The crying stopped a few minutes ago, or maybe it was longer than that. You have lost the ability to judge time accurately. Your eyes are open, but your vision has that strange, sandpapered quality that comes from a third or fourth interrupted night in a row, the kind where you never quite fell back into a real sleep, just a shallow gray drift between one waking and the next.

You are not just tired. You are a different version of yourself, running slower, feeling things more sharply and more distantly at the same time. Somewhere in the back of your mind, a voice is cataloging what needs to happen tomorrow: the early meeting, the permission slip, the pediatrician's appointment, the return call you keep forgetting to make. You register all of it and feel exactly none of it, because right now the only thing that matters is the silence coming from the other side of that door and whether it is going to hold.

You have probably read some version of the sleep advice. Maybe you have read a lot of it: the books, the forums, the long threads where other parents describe what worked for their child, which is rarely the same child as yours. Some of it contradicts itself. Some of it was clearly written by someone who forgot what this stage actually feels like from the inside, from the floor, at three in the morning, in the hallway, with one hand on the wall.

Here is what almost none of that advice tells you: you are not managing your child's sleep problem. You are managing a household sleep problem. The two nervous systems in that house, yours and your child's, are not running independently of each other. They are coupled. They are affecting each other in ways that are biologically real, measurable, and, once you understand them, genuinely actionable.

Your child wakes; your cortisol spikes. Your cortisol spike keeps you from returning to deep sleep even after your child settles. Your fragmented sleep shortens your fuse the next day. It makes your responses to your child's distress faster and less regulated, thereby heightening your child's own physiological arousal in the evening. A heightened child falls asleep harder. A child who falls asleep harder wakes more. And so on and so on, cycling through the nights, with both of you caught inside the same loop and neither of you the clear cause of it.

That loop is what this book is about. Not just your child's sleep, and not just your sleep, but the system the two of you are running together, and what it actually takes to change it from the inside.

You found this hallway. You are going to find the other side of it too.

Chapter Summary

This opening chapter does one essential thing before anything else in this book can be useful: it reframes children's sleep as inseparable from the sleep of the adults caring for them. Most sleep books for parents focus on the child as the problem to be solved and the parent as the one doing the solving, as though the parent exists outside the system, observing it with fresh eyes, a regulated nervous system, and plenty of reserve capacity. That is not the reality most parents are living. The reality is a coupled biological system, two or more nervous systems sharing a home, responding to each other's signals night after night, dragging each other upward toward better sleep or spiraling each other downward toward worse.

This chapter introduces the core framework that anchors everything that follows: the two-brain model. It is not a metaphor. The mechanisms are real; they operate through specific hormones, rhythms, and feedback loops that neuroscience has mapped with increasing precision over the past two decades. Understanding those mechanisms does not require a neuroscience degree. It requires only a willingness to see yourself as part of your child's sleep system, not outside it, and to accept that getting your own sleep under better control is one of the most powerful things you can do for your child's sleep. The rest of this book builds on that foundation, chapter by chapter, developmental window by developmental window.

What This Chapter Will Cover

- Why pediatric sleep advice feels contradictory. Different developmental windows respond to entirely different interventions, and most general advice does not specify which window it applies to. A strategy that transforms a six-month-old's nights may do nothing for a sixteen-month-old and actively backfire with a four-year-old. This book is organized by developmental window specifically to solve that problem, and this chapter explains the logic behind that structure.

- The coupled-system view of family sleep. When you track parent sleep and child sleep on the same chart for even a single week, patterns emerge that are invisible when you look at either one alone. Most parents discover that their own sleep timing, consistency, and fragmentation are more closely tied to their child's patterns than they expected. That discovery reframes where the leverage actually is.

- The science behind the two-brain model. The coupling between parent and child sleep is not just behavioral; it is hormonal and neurological. This chapter walks through the main biological mechanisms, including shared circadian entrainment, cortisol mirroring, oxytocin-mediated co-regulation, and how a child's developing nervous system actively uses a caregiver's regulated state as a template for its own regulation.

- Why sleep training is nervous-system training. For both parent and child. Any method that moves a child toward more consolidated, predictable sleep is simultaneously training the caregiver's nervous system to expect and trust that consolidation. The "training" is not just behavioral conditioning of the child; it is a parallel recalibration of the entire household's autonomic baseline.

• The role of sleep debt in parenting decisions. Research on decision-making under sleep deprivation is unambiguous: partial sleep loss degrades judgment, emotional regulation, and patience in ways that closely parallel the effects of low-level intoxication. This matters enormously for parents, because the decisions sleep-deprived parents make about their child's sleep are being made by brains that are least equipped to make them well.

• How to use this book in real time. Most parents will read this book while already running on fragmented, insufficient sleep. This chapter offers a practical framework for extracting the most immediately actionable guidance from each chapter even before the deeper material has fully landed, so that tonight's bedtime can be slightly better than last night's even on the first read.

• What comes next, and why the order matters. The book follows the neurodevelopmental timeline of the human sleep system from birth through adolescence, and reading it in that order is not just for completeness. Each developmental window introduces new neurobiological machinery, and understanding what changed and why is what makes the interventions make sense rather than feel like things you are trying to do.

The Body Clock You Never Knew You Were Sharing

Every living thing with a nervous system runs on a biological clock. In humans, that clock is housed in a tiny region of the hypothalamus called the suprachiasmatic nucleus, and it drives an approximately 24-hour rhythm that orchestrates not just sleep and wakefulness but body temperature, hormone release, digestion, immune function, and dozens of other physiological processes. You have one. Your child has one. And from the very first days of your child's life, those two clocks have been doing something easy to miss: they have been talking to each other.

The mechanism works through light, sound, feeding times, temperature, and, most powerfully, social cues. Human circadian rhythms are not sealed, self-contained clocks. They are open systems, continuously recalibrated by the environments in which they run. For an infant whose own clock is still immature and disorganized, the most powerful environmental calibrator available is the caregiver. The rhythmic predictability of the caregiver's behavior- when they appear, when they feed, when the lights go on, when the voice softens, and the movement slows- functions as an external timekeeper for the infant's developing system.

Diagram 1.1: Synchronized Circadian Clocks

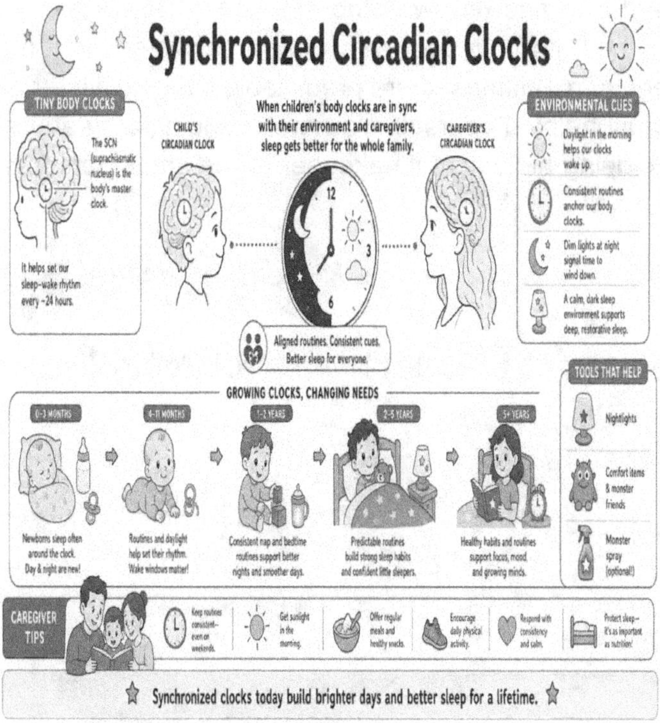

This is not a one-way transfer. While the caregiver provides rhythmic cues that help organize the infant's clock, the infant's signals simultaneously reorganize the caregiver's. New parents typically experience a dramatic disruption of their own circadian rhythms in the first months, partly from sleep fragmentation and partly from the constant biological demands of caring for a small person whose own rhythms are still taking shape. The two clocks are not just coexisting; they are actively co-shaping each other through a process researchers call entrainment coupling.

What this means in practical terms is that inconsistency in the caregiver's daily rhythms- irregular wake times, variable mealtimes, unpredictable social cues- carries a real cost for the infant's developing system. It is not a moral failing. It is a biological reality. The clock in your child's brain is partly reading yours. When yours is erratic, the signal it sends is erratic, and your child's clock has to work harder to find its own stability. Conversely, when you build more rhythm into your own days- more consistent light exposure, more regular mealtimes, more predictable evening wind-down- you are not just helping yourself. You are providing a clearer signal to the developing biological timekeeper in the small person sleeping down the hall.

Diagram 1.2: The Coupled Parent-Child Sleep System

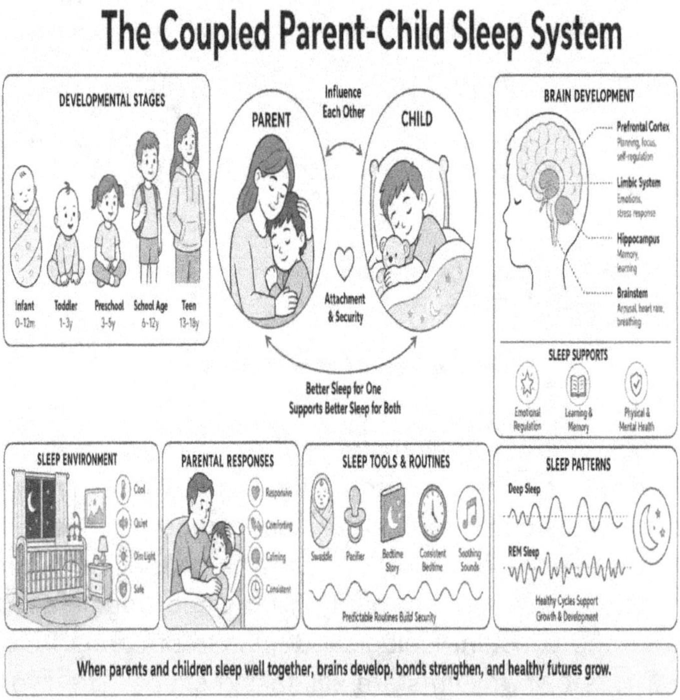

The ultradian rhythm is a shorter cycle nested inside the longer circadian one, and it matters enormously for understanding nighttime wake-ups. Adults cycle through sleep stages roughly every 90 minutes, moving from lighter to deeper sleep and back again. Children cycle faster: infants cycle every 45 to 60 minutes, toddlers gradually lengthening toward the adult pattern over the first several years. Each cycle ends with a brief lightening of sleep, a moment of partial arousal that most adults pass through without fully waking. Infants pass through the same moment, but their not-yet-mature regulatory systems are less able to handle the transition independently. They surface. They signal. And the caregiver, whose own cycle has been primed by months or years of responding, surfaces right along with them.

This is not a character flaw or a sign that something went wrong. It is the biological system doing exactly what it evolved to do. The infant signals; the attuned caregiver responds. The caregiver's arousal protects the infant. What the system was not evolved for is the modern context of full-time employment, sleep debt, and the expectation that the same adults managing all of this will also be functional, patient, and emotionally available the following day. The mismatch between the biological system and the modern demands placed on that system is exactly where the suffering lives, and it is exactly what this book is designed to help address.

When One Person's Bad Night Belongs to Everyone

You have probably had the experience of coming into a morning after a rough night and feeling it not just in your body but in the whole texture of the day. The threshold for frustration is lower. The things that normally roll off stick to you. The patience that usually flows fairly naturally has to be effortfully manufactured, and sometimes it cannot quite be manufactured fast enough. You know this feeling. What you may not know is that the people around you feel it too, not just as a behavioral change in you, but as a physiological response in themselves.

There is a substantial body of research on what is sometimes called emotional contagion, the rapid, largely unconscious transmission of emotional and physiological states between individuals in proximity. In parent-child relationships, this transmission is particularly efficient. Children, especially young children, do not just observe a parent's emotional state; their own nervous systems respond to it. A parent's elevated cortisol, communicated through vocal tone, facial expression, movement quality, and timing of responses, triggers a cortisol response in the child. A parent's calm, regulated state has the opposite effect. The child's nervous system is, in a very real sense, borrowing regulation from the caregiver's, using it as a scaffold while its own self-regulatory capacity is still under construction.

Diagram 1.3: The Household Sleep Debt Cascade

This creates a propagation pathway for sleep debt that runs through the entire household. One parent has a bad night (and with a young child, the two adults in the house often have different bad nights on different days, staggering their fragmentation in a way that leaves no one to recover). That parent wakes with elevated cortisol, blunted prefrontal function, and reduced capacity for emotional regulation. Their interactions with the child throughout the day show greater arousal, less predictability, and less calm attunement that the child's nervous system depends on. The child accumulates stress across the day, goes into the evening already more activated, has more difficulty settling at bedtime, wakes more overnight, and the cycle starts again.

Sleep researchers have documented the downstream effects of parental sleep loss on child behavior and development with increasing precision. What they find is not simply that tired parents are crankier, although that is true. They find that children whose caregivers are chronically sleep-deprived show measurable differences in cortisol patterns, emotional regulation, and sleep architecture itself. The household is a biological unit, not a collection of independent individuals. The sleep debt does not stay with the person who accrued it. It distributes.

Diagram 1.4: The Night-Waking Spiral

Understanding this cascade is not meant to add weight to an already heavy situation. It is meant to shift the target. If the problem is a system issue rather than a child's or a parent's, then the solution is also a system solution. You do not have to fix your child's sleep while somehow keeping your own sleep perfectly together. You do not have to be a better parent on less sleep. You have to start treating the system as a system, identifying the leverage points, and making changes that are sustainable for all the biological entities running in your house. That reframe is the whole point of this chapter and the whole foundation of this book.

The Cortisol-Oxytocin Loop at 3 A.M.

Cortisol and oxytocin are, in a simplified but useful sense, the two main characters in the nighttime story between parent and child. Cortisol is the activation hormone, the one that mobilizes energy and attention in response to a perceived challenge or threat. Oxytocin is the connection hormone, the one released during feeding, touch, eye contact, and responsive caregiving. In a well-rested parent, these two systems run in reasonable balance. In a sleep-deprived parent, that balance is significantly compromised.

The cortisol axis, technically the hypothalamic-pituitary-adrenal axis, is profoundly sensitive to sleep disruption. Even a single night of fragmented sleep raises baseline cortisol in the morning. It blunts the normal daytime pattern in which cortisol gradually declines from its morning peak through the afternoon and evening. The evening decline in cortisol is not incidental; it is part of what allows the brain to shift into sleep mode. When sleep deprivation keeps cortisol artificially elevated in the evening, both the parent and, through the transmission mechanisms described above, the child have more difficulty making that transition.

Diagram 1.5: Cortisol and Oxytocin Feedback Loop

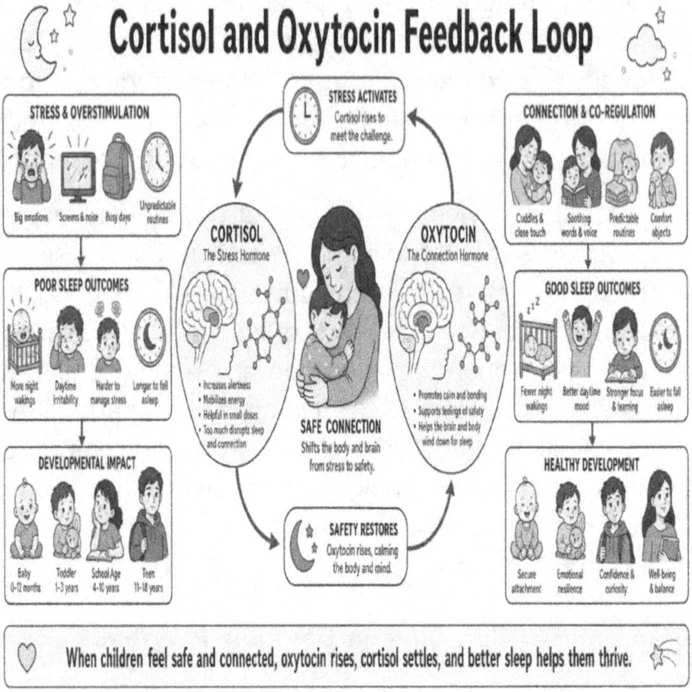

Oxytocin is both important and equally fragile. Responsive caregiving, including nighttime caregiving, is an oxytocin-generating activity. When you go to your infant at night, when you pick them up, and they settle against you, when you feed them or soothe them and feel their body relax into yours, your oxytocin rises. So does theirs. Oxytocin has a calming effect on the HPA axis; it literally damps down the cortisol response. This is part of why responsive caregiving feels so different from simply being awake in the middle of the night for no reason. The act of caring for someone you love has built-in biological compensation for the cost of being awake to do it.

The problem, and it is a significant one, is that oxytocin's capacity to buffer cortisol is limited, and under conditions of severe or chronic sleep deprivation, the balance tips. The HPA axis wins. Cortisol stays elevated. The caregiver begins to respond to the infant's nighttime signals not with the calm attunement that generates the oxytocin buffer, but with anxiety, frustration, or a desperate, resigned urgency that the infant reads clearly and that escalates rather than calms the infant's own arousal. This is not a failure of love. It is a failure of capacity, and capacity can be rebuilt with the right support and the right changes to the surrounding system.

Diagram 1.6: Co-Regulation Map

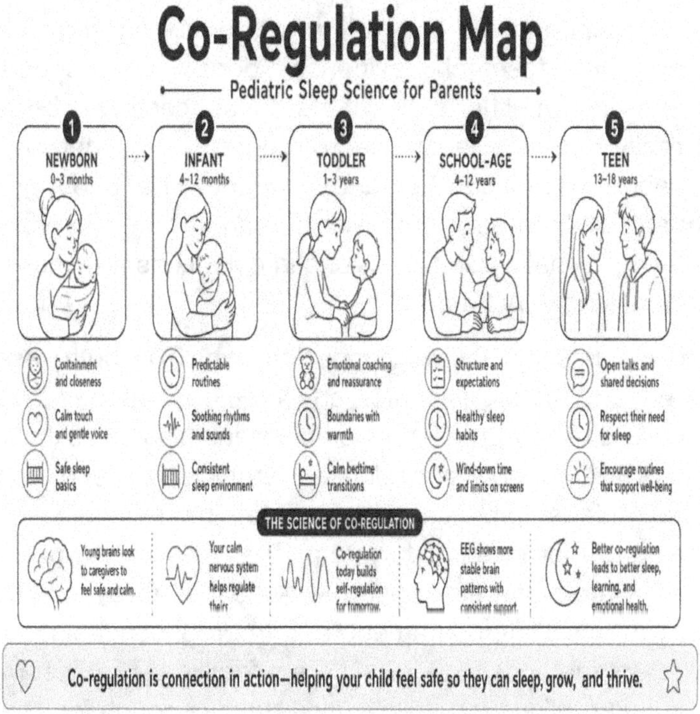

What makes the cortisol-oxytocin story clinically relevant is that it helps explain something parents often describe with guilt and confusion: the way they can be simultaneously deeply loving and acutely depleted, the way they can desperately want to comfort their child and simultaneously feel a flicker of something darker when the crying starts again at 2 A.M., the fourth or fifth time that night. That flicker is not a character defect. It is a depleted HPA axis sending a distress signal. Naming it helps. Understanding that the system can be changed helps more. And that is what the rest of this book is organized to deliver.

Sleep Training Is Nervous-System Training

The phrase "sleep training" has accumulated an enormous amount of cultural charge. It conjures up, depending on who you ask, either the responsible act of teaching a child an essential life skill or a form of infant abandonment that will produce lasting psychological damage. The research evidence does not support the catastrophizing on either end, but the debate continues because it is not really a debate about neuroscience. It is a debate about values, about what kind of parent you want to be, about guilt and identity and the competing pressures bearing down on exhausted adults trying to make good decisions in conditions that make good decision-making very hard.

This book takes the following position, which is both research-supported and compassionate: any method that helps a child develop more consolidated, predictable sleep also helps that child's caregiver develop more consolidated, predictable sleep, and both of those things produce measurable improvements in the functioning of the household system. The question is not whether to move toward better sleep but how to do it in a way that fits your values, your child's temperament, and the real constraints of your life. Different methods work for different families. The best method is the one you will actually follow consistently enough for it to work.

Diagram 1.7: The Sleep Engineer Framework

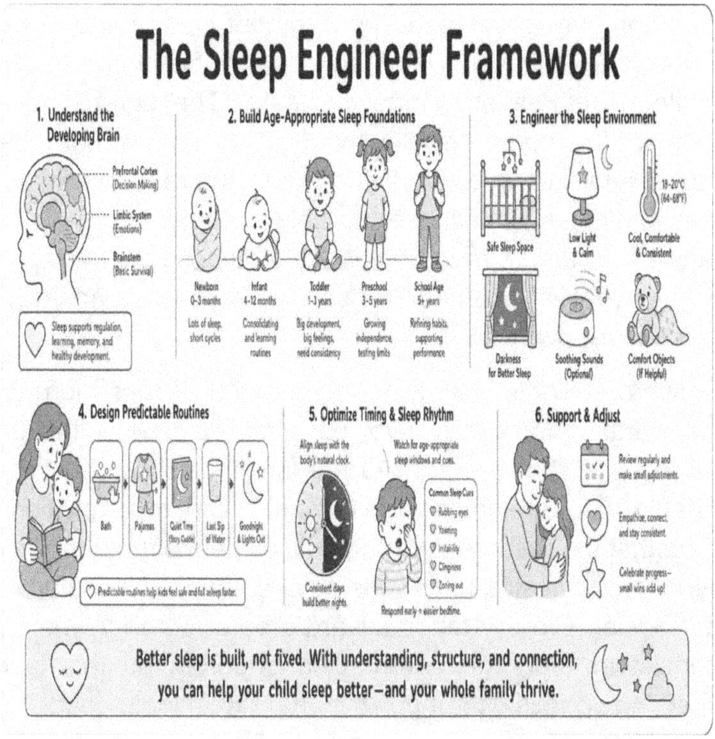

What rarely gets discussed in sleep-training conversations is the degree to which the process changes the parent's nervous system, not just the child's. Sleep architecture is not simply a matter of hours clocked. It is a sequence of stages, each performing specific biological work: slow wave sleep for physical restoration and memory consolidation, REM sleep for emotional processing and creative integration, light sleep as the gateway between the two. When fragmentation repeatedly interrupts that sequence, the deepest and most restorative stages are disproportionately lost because the brain prioritizes the lighter, shallower stages when it is uncertain whether uninterrupted sleep will be available. The result is a person who spends time in bed but wakes feeling like the sleep did not quite count.

When a child begins sleeping in longer stretches, the parent's sleep architecture begins to recover. Deep slow-wave sleep, which is disproportionately suppressed by fragmentation, starts to rebuild. Prefrontal cortex function, which is among the first things to degrade under sleep deprivation, starts to return. The parent wakes up slightly more functional, handles the day with slightly more reserve, and goes into the evening with slightly more capacity for the bedtime routine. And because the child's nervous system mirrors the parent's, a calmer, more regulated bedtime leads to a child who settles more easily, which further consolidates sleep and gives the parent more recovery time.

This upward spiral is real. It is documented. It takes longer than anyone wants it to, and it involves a middle period that can be difficult in ways that are hard to prepare for fully. What makes it sustainable is understanding that the difficulty of the transition is not evidence that the method is wrong. It is evidence that the system is adjusting, and adjustment periods have a shape. They start harder and get easier. Biology tends toward greater stability when inputs are consistent, and the parent who understands this is better equipped to persist through the harder middle part without abandoning the process prematurely. But it is the trajectory that the system moves toward when the right conditions are in place. Sleep training, in any of its forms, is the process of creating those conditions. Seen through that lens, it is less about making a child cry alone in the dark and more about gradually building a household nervous-system environment in which everyone, including the child, can find and maintain rest.

It is also worth saying something about the specific methods available, because the landscape of options is genuinely wide and the language around them is often more polarized than the evidence warrants. Gradual extinction (sometimes called the Ferber method or controlled crying) involves the parent checking in at increasing intervals, allowing the child to experience brief periods of settling independently while still receiving reassurance. Full extinction (often called cry-it-out) involves allowing the child to settle without parental intervention after the initial bedtime. Both have robust evidence for effectiveness in healthy infants, and both have been studied for adverse outcomes, with repeated large studies finding no lasting psychological harm. The no-cry approach, most famously associated with Elizabeth Pantley's work, uses very gradual adjustments to sleep associations and is slower but more comfortable for parents who find other approaches distressing. The chair method, also called sleep fading or gradual withdrawal, involves the parent staying present but progressively less involved—all of these work. The differences in outcomes across methods are much smaller than the debates around them suggest. What predicts success most consistently is parental follow-through, which is determined far more by fit with the parent's values and capacity than by the method's theoretical elegance.

The nervous-system framing also reframes what it means when sleep training does not immediately work. If the intervention is not just behavioral conditioning of the child but a recalibration of the whole household's autonomic baseline, then setbacks are not evidence that the child is broken or the method is wrong. They are evidence that the system is recalibrating and that the new pattern has not yet become the stable attractor it will eventually be. Patience with the process is easier when you understand what the process actually is.

Diagram 1.8: The Tired Parent Decision Tree

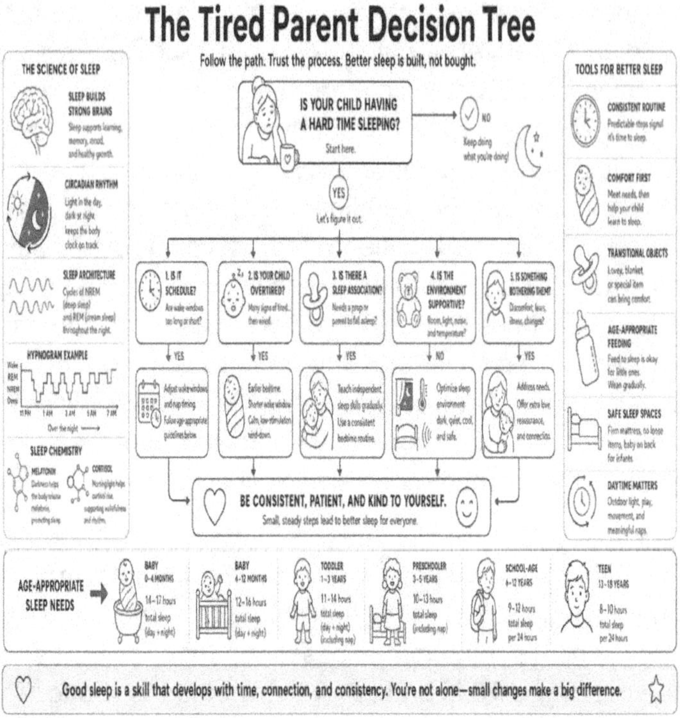

There is also something important to say about the parent who cannot do sleep training, whether because of the child's temperament, medical needs, their own mental health, their living situation, or simply a deep and considered conviction that it is not right for them. This book does not require sleep training. What it requires is a willingness to engage with the household system as a system, to identify what is creating fragmentation and what could reduce it, and to make changes that are sustainable for your particular family. For some families, that will involve structured sleep training. For others, it will involve graduated changes to the environment, the schedule, and the caregiver's own sleep habits. For all families, it will involve understanding the two-brain model well enough to use it.

Two Brains, One Rhythm

The phrase "two brains, one rhythm" is a compression of everything this chapter has covered, and it is the lens through which every subsequent chapter in this book is written. It means, concretely, that the biological rhythms of a child and a caregiver in the same household are not independent variables. They are coupled. They influence each other continuously through hormonal signals, behavioral cues, shared environmental timing, and the deep mammalian biology of social bonding. Treating them as independent, trying to fix one without attending to the other, is like trying to fix the temperature in one room of a house without considering the HVAC system the whole house shares.

The model has several specific implications for how you read and use this book. First, every recommendation about what to do for your child at a given developmental stage comes with an implicit parallel recommendation about what the caregiver needs to be doing to support it. That parallel is sometimes stated explicitly, sometimes embedded in the reasoning. Pay attention to it. Second, the model means that there is no such thing as a purely behavioral sleep intervention for a child. Every intervention changes the caregiver's behavior and physiology too, and those changes feed back into the child's system. This is good news: it means there are more leverage points than child-sleep advice usually acknowledges.

Diagram 1.9: Two Brains One Rhythm

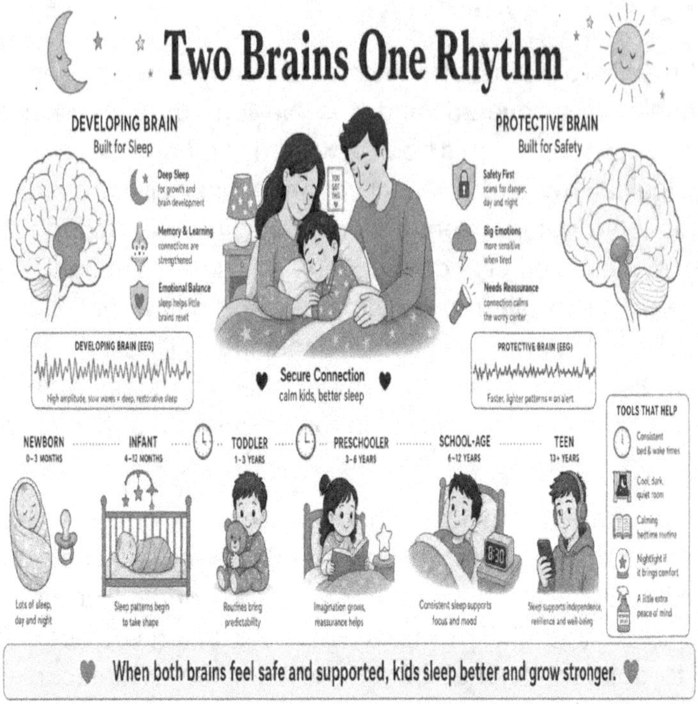

There is a fourth, easy-to-overlook implication: the two-brain model reframes the question of when to start. Parents often ask whether it is too early or too late to address their child's sleep, whether they have missed a window or are imposing a change at the wrong developmental moment. The honest answer is that there is no single universally right moment. Still, there are windows within each developmental stage where change is relatively easier, and the chapters ahead will identify those windows precisely. The more important truth is that any moment when you are ready and able to work on the system is a valid moment, because the system is always running, always influencing itself, always capable of moving in a better direction given the right inputs.

Third, and perhaps most important, the two-brain model is a model of compassion. It means that the difficulty you are having right now is not a sign of failure. It is a sign that two biological systems are caught in a pattern that is not serving either of them, and that patterns can change. The same biological coupling that creates the problem is the mechanism through which the solution works. Your child's nervous system is exquisitely sensitive to yours. That sensitivity is what makes the loop hard to break when you are both exhausted, but it is also what makes the changes you make so powerful. When you get more rest, your child's nervous system knows. When you regulate your own activation more effectively, your child's settles more easily. The two-brain model is not just a description of why things are hard. It is a description of why the work you do on your own sleep and nervous-system health is some of the most effective parenting work you can do.

The remainder of this book is organized to make that work practical, chapter by chapter, developmental window by developmental window. Each chapter covers a specific stage: the newborn period, the four-month transition, infancy, toddlerhood, the preschool years, early school age, and adolescence. Each chapter introduces the neurobiology specific to that window, because the biology changes dramatically and the interventions that work change with it. A toddler's resistance to bedtime is not the same biological phenomenon as a newborn's fragmented sleep or a teenager's phase shift. They require different responses, and those responses make much more sense when you understand why the biology is doing what it is doing.

Throughout all of those chapters, the two-brain model runs as a constant thread. Whatever developmental window you are in, whatever specific challenge you are facing, the question is always the same: what is happening in both nervous systems, how are they affecting each other, and what can change in the system to move both of them toward more rest? That question is what makes this approach an engineering approach rather than a prescriptive one. You are not following a set of rules. You are understanding a system and making targeted interventions. The difference matters because your child is not generic, and your family is not generic. The principles are universal. The application is yours.

You are already doing this work. The fact that you are reading this, on whatever amount of sleep you managed last night, means you are paying attention to the system and trying to understand it. That is the beginning. Everything else in this book is the next step.

Try This Tonight

For this age range: 0 to 18 years

1. Before you do anything else, write down your own actual sleep timing for last night: what time you got into bed, what time you think you fell asleep, how many times you woke, and what time you finally got up. Do not estimate generously. Write what actually happened. This is your baseline, and you cannot improve a baseline you have not honestly named.

2. Do the same for your child: approximate bedtime, how long it took them to settle, how many times they woke, and what time the day started. Put both records on the same piece of paper or the same note on your phone. Look at them side by side. Notice whether the patterns resemble each other more than you expected.

3. Pick one fixed anchor point for tomorrow morning and commit to it tonight. An anchor point is a consistent wake time that you protect regardless of how the night went. It does not have to be early. It has to be consistent. If your child is old enough to have a morning anchor, pick one for them as well, even if it's a loose one. Consistent morning anchors are the single most effective tool for stabilizing a disrupted circadian system in both adults and children.

4. Look at your evening environment in the two hours before your target bedtime. How much bright light is present? How much screen activity? How much stimulating conversation or activity? Make one small change tonight: dim one light source

or move one screen activity earlier. You are not redesigning the whole evening. You are making one small change to the signal your brain receives about the time.

5. If your child is old enough to understand simple instructions, tell them one thing about tonight that will be the same as last night and the same as tomorrow night. It does not have to be elaborate: the same stuffed animal, the same two songs, the same lamp on low. Predictability is what the developing nervous system uses to build its sense of what comes next, and in the child's brain, that next thing is sleep.

6. Before you go to bed, take two minutes to do something that has no task attached to it. Sit outside if it is warm. Stretch. Make tea and drink it without looking at your phone. The transition from "doing" to "resting" is not automatic for the adult nervous system under stress, and two intentional minutes of transition signal to your cortisol system that the day is actually over. This is not a luxury. It is a physiological necessity, and two minutes is enough to begin the signal.

7. If you woke up last night and stayed awake for a long period, notice whether you spent that time on a bright screen. If you did, tonight try keeping the room as dark as possible when you respond to your child. Use a dim nightlight at most. The light signal is the most powerful input to the circadian clock, and even brief bright light exposure in the middle of the night can delay your next sleep cycle by thirty minutes or more.

Parent Takeaways

- Your sleep and your child's sleep are one coupled biological system; improving one without attending to the other is working with one hand tied behind your back.
- The hormonal connection between parent and child is real and measurable: your cortisol state transmits to your child's

nervous system, and your calm is one of the most effective sleep tools you have.

• Sleep training, in any form, is nervous-system recalibration for the whole household, not just behavioral conditioning of the child.

• Sleep deprivation degrades the exact capacities (patience, emotional regulation, sound judgment) that good nighttime parenting requires; protecting your own sleep is not selfish, it is structural.

• The best sleep method for your family is the one you will actually follow consistently; this book will help you find what that is for your specific child, your specific stage, and your specific life.

• Every developmental window has its own neuroscience and its own leverage points; matching the intervention to the window is the highest-impact move available to you, and the rest of this book is organized to make that matching possible.

2 The Newborn Brain: First 3 Months

It is 2:47 in the morning, and you are standing in the nursery for the fourth time tonight, swaying side to side with a three-week-old on your shoulder. The baby is not screaming anymore, but she is not asleep either. She is wide awake, staring at the ceiling fan with the focused intensity of someone who has never seen a ceiling fan before, which, to be fair, she has not. You lowered the lights at 7 p.m. You put on the white noise machine. You did the bath, the swaddle, the gentle bounce. You did everything every blog post and well-meaning family member told you to do. And still she is here, at 2:47 a.m., fascinated by the ceiling fan.

Here is what nobody told you: she has no idea it is nighttime. Not because she is stubborn, and not because you are doing anything wrong. Her brain does not have the biological machinery yet to tell day from night. The part of the brain responsible for tracking time, a tiny cluster of neurons called the suprachiasmatic nucleus, is present at birth but has not yet learned to listen to the world's light-dark signals. She is not on a 24-hour clock. She is on no clock at all.

Take a breath. This is not a parenting failure. This is developmental neurology playing out exactly as it should.

You are in the fourth trimester. Your baby spent nine months in an environment with no sunlight, no silence, and no such thing as nighttime. She slept when she was warm and fed and heard the muffled rhythm of your heartbeat. She woke when she was hungry or uncomfortable. The outside world is brand new to every single one of her systems, and her brain is doing the enormous work of figuring it all out. Sleep, for her, is not a behavior you shape yet. It is a biological process still under construction.

What gets so many new parents is the gap between expectation and reality. Someone will tell you a story about a friend's baby who "slept through" at six weeks, and you will wonder what you are doing wrong. The answer is nothing. You are doing nothing wrong. That baby was likely sleeping one long stretch in the early evening, which is not the same as the unbroken eight-hour night adults dream about. And even if it were, it would be the exception so far outside the bell curve that building an expectation around it is like expecting your newborn to walk because you heard about one kid who took steps at seven months.

You are not behind. Your baby is not broken. The first twelve weeks are supposed to look like this, and by the time you finish this chapter, you are going to understand exactly why, and that understanding is going to make the 3 a.m. ceiling-fan moments feel a little less desperate.

Chapter Summary

This chapter is about what is actually happening inside your newborn's brain while you are trying to keep up with her around the clock. Newborns sleep a lot, somewhere between 14 and 17 hours every day, but they do it in short, scattered bursts because their internal clock has not yet synced to the 24-hour world. Their sleep cycles are short, roughly 50 minutes compared to the 90-minute cycles adults run, and the sleep itself looks completely different from what you experience at night. The chemical that makes us feel sleepy in the evening, melatonin, does not begin to flow in a reliable rhythm until around 8 to 12 weeks of life.

Understanding this biology will not magically get you more sleep tonight. But it will almost certainly reduce the anxiety that piles on top of the exhaustion, because anxiety costs energy you cannot afford right now. When you know that night waking is not a flaw to be fixed but a neurological reality to be waited out, you stop fighting biology and start working with it. This chapter also covers what safe sleep actually looks like at this age, how feeding and sleep are tangled together in completely normal ways, and how to protect at least some of your own sleep while riding out these first weeks.

What This Chapter Will Cover

- The unfinished circadian clock. The suprachiasmatic nucleus drives our sense of day and night, but in a newborn it has not yet learned to respond to light and darkness. You will understand why this happens, how long it typically takes to change, and what you can do to gently nudge the process along without adding pressure to yourself or the baby.

- Active Sleep and Quiet Sleep. Newborns do not cycle through REM and NREM the way adults do. They have their own versions, called Active Sleep and Quiet Sleep, and the proportion of each tells us something important about how the brain is developing. Knowing what each one looks like will also stop you from picking up a sleeping baby who seems restless.

- The 50-minute sleep cycle. Newborns complete a full sleep cycle in about 50 minutes, compared to 90 minutes in adults. Understanding this rhythm helps you predict when the baby is likely to stir and what to do, and what not to do, in those transition moments.

- Why sleeping through the night is not possible yet. The phrase "sleeping through the night" is one of the most harmful pieces of language in parenting culture. This section explains the biology that makes full-night sleep impossible during the first 12 to 16 weeks and gives you a more realistic set of milestones to look forward to.

- The fourth-trimester framework. Your baby is not a small adult who needs to learn better habits. She is finishing a developmental process that birth interrupted. This framework recalibrates everything and makes the whole first trimester feel more survivable.

- Safe sleep, explained. The American Academy of Pediatrics guidelines are not bureaucratic rules; they are the product of decades of research into what actually keeps babies safer at night. This section explains the logic behind back-to-sleep, firm flat surfaces, room-sharing without bed-sharing, pacifier use, and the swaddling cutoff, so you follow the guidelines because you understand them, not just because someone told you to.

- Feeding, hunger, and sleep. At this age, sleep and feeding are almost the same thing. Hunger is the primary reason newborns wake, and cluster feeding at certain times of day is normal, expected, and not a sign that your supply is failing. Understanding this cycle removes a huge source of nighttime anxiety.

The Clock That Has Not Started Yet

There is a region deep in your brain called the suprachiasmatic nucleus, or SCN. It sits in the hypothalamus, just above where the two optic nerves cross. Light information travels from your eyes to the SCN every day, and the SCN uses that information to set and maintain your 24-hour biological clock. It tells your body when to release melatonin, when to raise your core temperature, when to be alert, and when to wind down for sleep. Your SCN is so well calibrated that even small shifts in your light exposure, like crossing time zones or staying up late on your phone, can knock it off track.

Your newborn's SCN is present at birth. The neurons are there, the structure is there. What is not there yet is the connection between the SCN and the environmental cues that train it. In adults, that connection is strong and well-established. In a newborn, it is fragile and still forming. Her SCN is essentially waiting to be calibrated, and that calibration happens gradually over the first several weeks of life as she accumulates light-dark exposure in the outside world.

Diagram 2.1: The Unfinished SCN

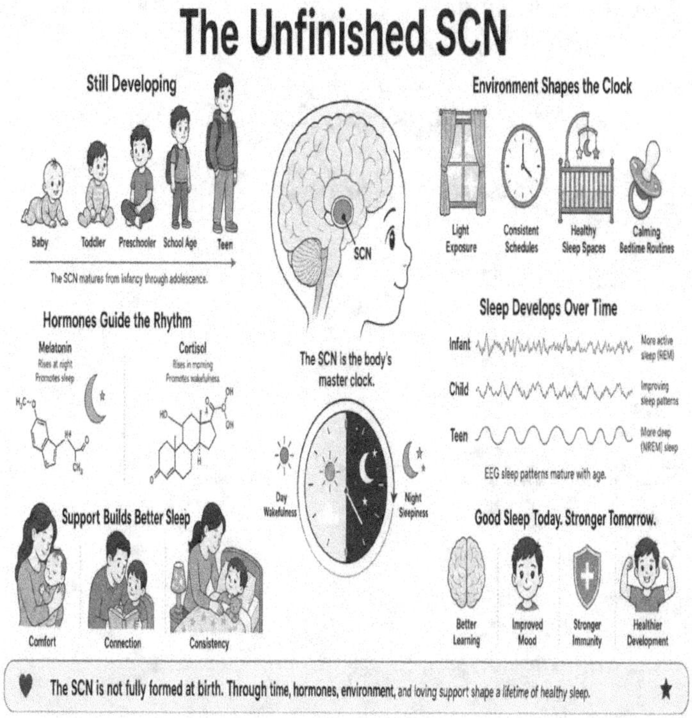

Before that calibration happens, she has no physiological anchor to the 24-hour day. Her sleep and wake periods are distributed roughly evenly across the 24 hours. You will almost certainly notice that she seems most wakeful in the late evening and early morning hours. This is partly because many babies emerge from the womb already biased toward late-night wakefulness, possibly because fetal movement tends to peak in the evening when the mother is resting. The gentle rocking of walking has stopped. She spent months being most active at night, and that rhythm does not reverse instantly.

The practical upshot is this: in the first four to six weeks, you cannot schedule a newborn, and trying to do so will exhaust you without producing results. The goal is not a schedule. The goal is to begin providing the environmental cues that will eventually help her SCN calibrate. That means bright light and gentle activity during daytime, and dim light, quiet voices, and minimal stimulation during nighttime feeds. You are planting seeds. They will not bloom for several weeks, but they will bloom.

Diagram 2.2: Day-Night Confusion Timeline

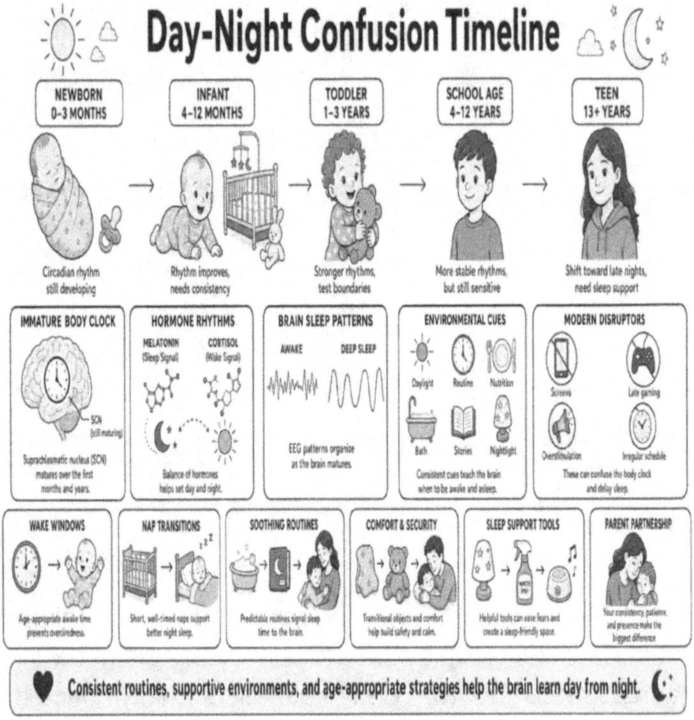

Around 8 to 12 weeks, something starts to shift. This is when the pineal gland begins producing melatonin in a rhythmic pattern, rising in the evening and falling in the morning. You will likely notice this before you understand what is causing it: the baby starts to have one slightly longer stretch of sleep in the first part of the night. She may begin to show a clearer "fussy evening" followed by a longer initial sleep. This is the SCN coming online, slowly, and it is one of the more encouraging milestones of the fourth trimester. You do not need to do anything to make it happen. You need to keep providing consistent cues and wait.

Active Sleep, Quiet Sleep, and the Developing Brain

If you have watched a newborn sleep, you have probably noticed how active the whole process looks. Her eyes flutter under closed lids. Her face twitches and scrunches. Her arms make tiny jerking movements. Her breathing is irregular, sometimes fast, sometimes slower, occasionally punctuated by a pause that makes you lean in to check. This is Active Sleep, and it is completely normal.

Active Sleep is the newborn equivalent of REM (rapid eye movement) sleep. In adults, REM is the stage associated with vivid dreaming, emotional processing, and memory consolidation. In newborns, Active Sleep is thought to serve a different but equally important function: it is when a massive amount of brain development happens. During Active Sleep, the brain generates internal signals that strengthen neural connections, lay down the foundations for sensory processing, and build the infrastructure for everything from vision to language. A newborn who spends a lot of time in Active Sleep is not a poor sleeper. She is doing important neurological work.

Diagram 2.3: Active Sleep vs Quiet Sleep

Quiet Sleep is the newborn equivalent of NREM sleep. During Quiet Sleep, your baby will look much more like a sleeping adult: still, with regular breathing, no eye movements, relaxed face and hands. This is the deep, restorative phase, and it is shorter and less frequent in newborns than in adults. In the first few weeks, Active Sleep accounts for roughly 50 percent of total sleep time, compared to about 20 to 25 percent in healthy adults. As the weeks pass and the brain matures, Quiet Sleep becomes longer and more frequent, and the ratio shifts toward the adult pattern. By about six months, the architecture starts to look significantly more grown-up.

One of the most practical takeaways from this is that not all movement during sleep means your baby needs you. Active Sleep looks uncomfortable, but she is not. If you respond to every twitch and flutter by picking her up, you may interrupt a sleep cycle she was about to complete on her own. Learning to wait a moment before responding, to give her 30 to 60 seconds to see if she settles, will serve you well from these first weeks all the way through toddlerhood.

The 50-Minute Newborn Sleep Cycle

Adult sleep cycles run roughly 90 minutes from start to finish. You fall through lighter NREM stages into deep slow-wave sleep and then back up through lighter sleep into a REM period, and that whole arc takes about an hour and a half. Because the transitions between cycles are the lightest moments of sleep, adults often briefly surface to a near-waking state every 90 minutes or so. Most adults do not notice because they roll over and fall back into the next cycle without fully waking. But if something disrupts that transition moment- a noise, a discomfort, an unfamiliar feeling- waking becomes more likely.

Newborns run shorter cycles, approximately 50 minutes from start to finish. That means they reach a transition point more frequently and have more opportunities to wake between cycles fully. This is not a design flaw. Short cycles with high Active Sleep content appear to serve the developmental needs of a brain that is growing at a pace it will never match again in life. But it does explain why a newborn who seemed deeply asleep 20 minutes ago is suddenly fussing again.

Diagram 2.4: Newborn 50-Minute Sleep Cycle

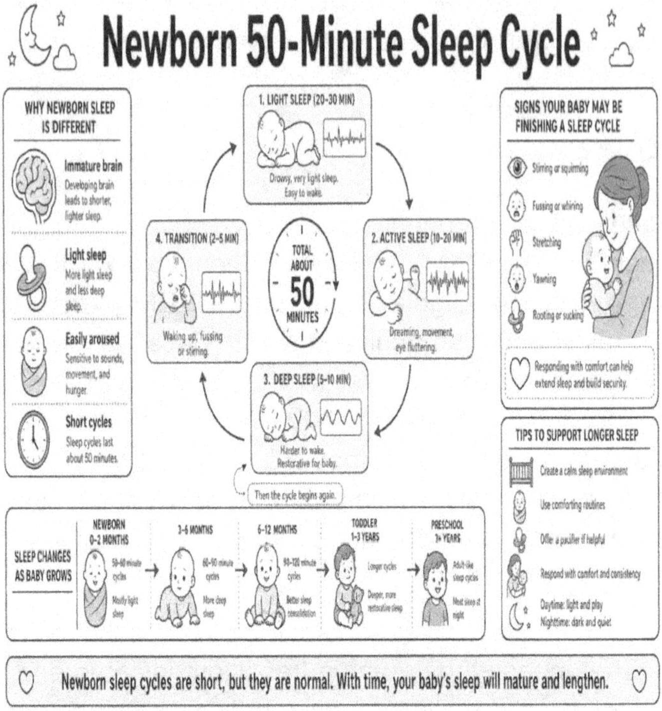

The cycle also explains one of the most common new-parent frustrations: the baby who falls asleep in your arms and then wakes the moment you put her down. What is happening is a transition between sleep stages. She was in Active Sleep, which involves some degree of muscle tone and sensory awareness. As you transfer her, she senses the change in temperature, position, and the loss of your heartbeat and breath sounds. If she is at a light point in her cycle, that sensory shift is enough to tip her from sleep to waking.

There are a few strategies that help. One is what is sometimes called "drowsy but awake," where you put the baby down before she is fully asleep so she learns to complete the transition into deeper sleep in the place where she will spend the rest of the night. This is genuinely useful, though at two or three weeks it may not work reliably, and that is fine. You are introducing the concept, not demanding the skill. Another strategy is the "wait and warm" approach: let your body heat warm the mattress surface for 30 to 60 seconds before lifting her away, so the temperature shift is less jarring. Some parents find that keeping one hand on the baby's chest for a moment after laying her down helps bridge the transition.

None of these are magic. Some nights they work beautifully. Some nights nothing works and you hold her. That is also fine.

Why Sleeping Through the Night Is Not Possible Yet

Let us talk honestly about the phrase "sleeping through the night," because it does more harm than almost any other piece of infant sleep language. Parents hear it constantly, from family members, from pediatric nurses at discharge, sometimes from well-meaning friends whose babies happened to sleep longer. And it becomes this looming benchmark, something you are supposed to be working toward and achieving, a measure of your competence and the baby's health.

Here is the biological reality. A full eight-hour stretch of unbroken night sleep requires a functioning circadian rhythm, a consolidated sleep architecture with long NREM periods, and a stomach capacity large enough to go several hours between feeds. Newborns in the first 12 to 16 weeks have none of these things. The circadian rhythm is not established. The sleep architecture is dominated by short Active Sleep cycles. And the stomach capacity of a newborn is genuinely tiny. In the first days of life, a newborn's stomach holds roughly 5 to 7 milliliters. By two weeks, it has expanded to around 60-90 milliliters, which is still a very small amount. Breastfed babies digest milk faster than formula-fed babies, meaning they reach hunger again sooner.

Diagram 2.5: Fourth-Trimester Expectations Chart

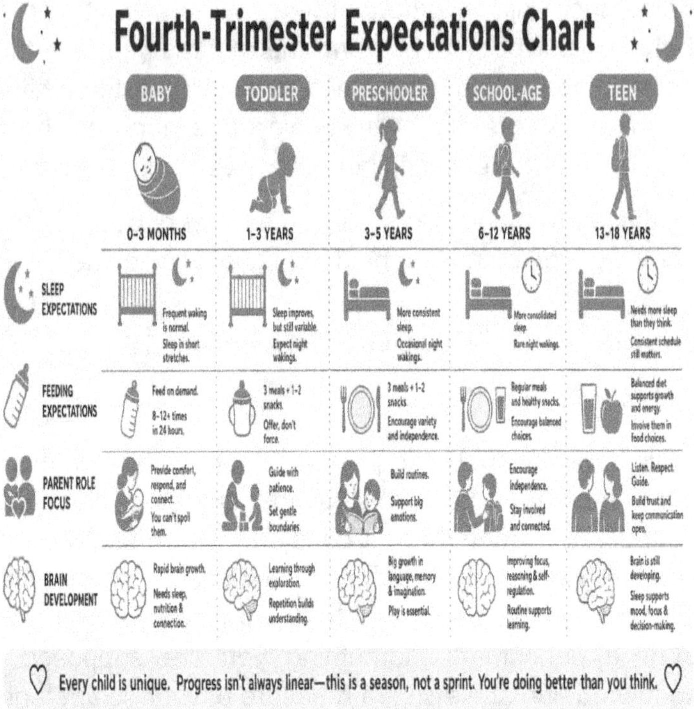

This is not a problem to be solved with more food before bed, or a stricter schedule, or any technique you find in a sleep training book. Night waking in this age window is the appropriate biological response of a healthy baby to the physical conditions of early infancy. It is protective in nature. Babies who sleep too deeply too early in life are at increased risk, which is part of why the relationship between arousal from sleep and safe sleep guidelines is so Important. Frequent waking is not a sign that something is wrong. It is a sign that normal developmental processes are intact.

A more realistic benchmark for this window is a single longer stretch of three to four hours sometime in the night, typically in the early part, followed by shorter periods of one to two hours. If you get that, you are doing well. If you are consistently getting two-hour stretches, you are in the normal range. Most parents find that somewhere between 10 and 16 weeks, they notice the first stretch beginning to lengthen. It rarely happens linearly; there will be regressions around growth spurts and developmental leaps. But the overall trajectory is toward longer stretches, and it will come.

The Fourth Trimester

The concept of the fourth trimester has been around in developmental circles for decades. Still, it entered mainstream parenting language more recently, and it is genuinely one of the most useful frames for surviving the first three months. The idea is simple: human babies are born earlier in their development than most mammals because of the size of the human head relative to the birth canal. A newborn foal can stand within hours of birth. A newborn human cannot sit up unassisted for four to six months. We are, in a very real developmental sense, born about three months too early.

The fourth trimester is the name for that external womb period. Your baby spent nine months in an environment characterized by constant warmth, constant gentle motion, constant noise at a moderate level (the whoosh of blood through vessels, your heartbeat, muffled external sounds), continuous access to nutrition, and no such thing as a day-night cycle. The outside world is a radical departure from all of that. The goal of the fourth trimester is not to train your baby out of needing comfort. It is to provide the womb-like conditions her nervous system still expects while she gradually adjusts to the outside world.

Diagram 2.6: Newborn Hunger and Sleep Loop

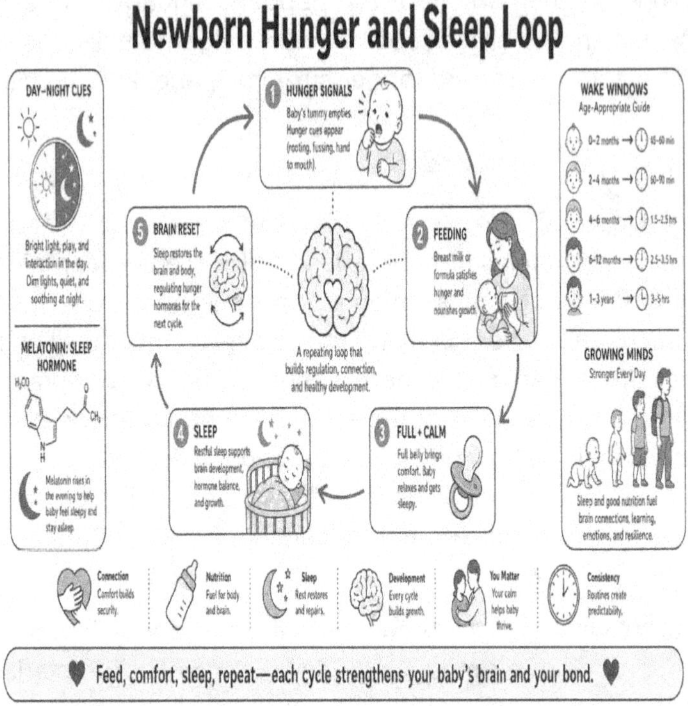

This framing changes the question from "how do I get my baby to sleep better?" to "how do I recreate the conditions her nervous system is looking for?" The answers include swaddling, which mimics the contained feeling of the uterus. White noise at moderate volume (about 65 decibels, roughly the level of a shower), which mimics the ambient sounds she heard in utero. Being held and carried, which provides the motion and heartbeat contact she is biologically primed to expect. None of these are habits that will doom you to a toddler who cannot sleep without being rocked. They are appropriate, evidence-based responses to a developmental stage, and most babies gradually need them less as the fourth trimester ends and their nervous systems mature.

The fourth-trimester mindset also permits you to stop trying to optimize. You do not need to have a schedule at three weeks. You do not need to do sleep training at six weeks. You do not need to have your baby sleeping independently in two-hour naps before returning to work. What you need is to keep the baby fed, warm, and safe, to get through each day, and to protect whatever scraps of sleep you can find for yourself. Everything else is secondary.

Safe Sleep: What It Is and Why It Matters

Safe sleep is not a set of arbitrary rules. It is a set of guidelines built on decades of research into sudden infant death and sleep-related infant deaths, and following those guidelines genuinely and meaningfully reduces risk. The American Academy of Pediatrics updates these guidelines as new evidence emerges, and the current guidance is clear: every sleep, every time, follows the same standards.

Back to sleep means placing your baby on her back for every nap and every overnight sleep until she can roll both ways independently. Babies who can roll from back to front and front to back on their own can be left in whatever position they roll into, but until that milestone, every placement is on the back. This recommendation is not about comfort; many babies show mild signs of preferring their sides or stomachs, particularly those with some reflux. The guidance holds regardless, because the research on arousal thresholds and airway protection is that consistent.

Diagram 2.7: Safe Sleep Setup (AAP-Aligned)

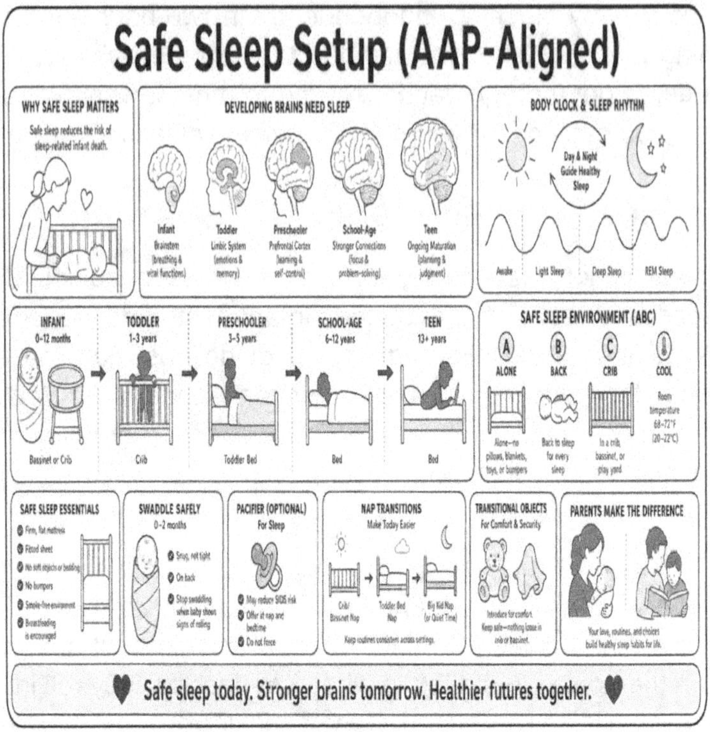

The sleep surface matters enormously. It should be firm and flat, covered by a fitted sheet designed for that specific surface, with nothing else in the sleep space. No blankets, no pillows, no bumpers, no positioners, no stuffed animals, no sleep wedges, no incline sleepers. The reason for all of these is the same: soft or loose materials around an infant create a suffocation and entrapment risk that is not visible until something goes wrong. The sleep space should look almost bare to adult eyes, and that is exactly right.

Room-sharing without bed-sharing is one of the more nuanced AAP recommendations and one of the most important. Sharing a room with your baby for at least the first six months, and ideally the first year, is associated with a meaningful reduction in sleep-related death risk. The protective mechanism is thought to involve heightened sensory arousal in response to the proximity of a breathing adult. Bed-sharing, on the other hand, is associated with increased risk, particularly on soft sleep surfaces, with extremely fatigued adults, or in situations involving alcohol or sedating medications. The recommendation is a bedside bassinet or separate sleep surface in the same room, not a shared bed.

Pacifiers have a documented protective association with safe sleep outcomes when given after breastfeeding is established, typically around three to four weeks. If your baby takes a pacifier and it falls out after she is asleep, you do not need to replace it. If she is breastfeeding and you want to try a pacifier, the usual guidance is to wait until the latch is comfortable and consistent. Some breastfed babies accept pacifiers without any disruption to nursing; others show some nipple confusion, and there is no reliable way to predict which category your baby will fall into.

Swaddling: Comfort With a Cutoff

Swaddling is one of the oldest infant soothing techniques in human history, and research generally supports it: a well-executed swaddle reduces startling, extends sleep bouts in the first weeks, and provides a contained, snug feeling that mirrors the intrauterine environment. For many newborns, it is genuinely one of the most effective tools in the fourth-trimester toolkit.

The technique matters. The swaddle should be snug around the arms and torso, but with enough room at the hips and legs for the knees to rise and the hips to spread slightly. Swaddling the hips tightly in a straightened position is associated with developmental dysplasia of the hip, a real risk that is easily avoided by ensuring the lower half of the swaddle has some flex. The classic hip-healthy swaddle allows the legs to frog up naturally while keeping the arms close to the body.

Diagram 2.8: Swaddling Safety Window

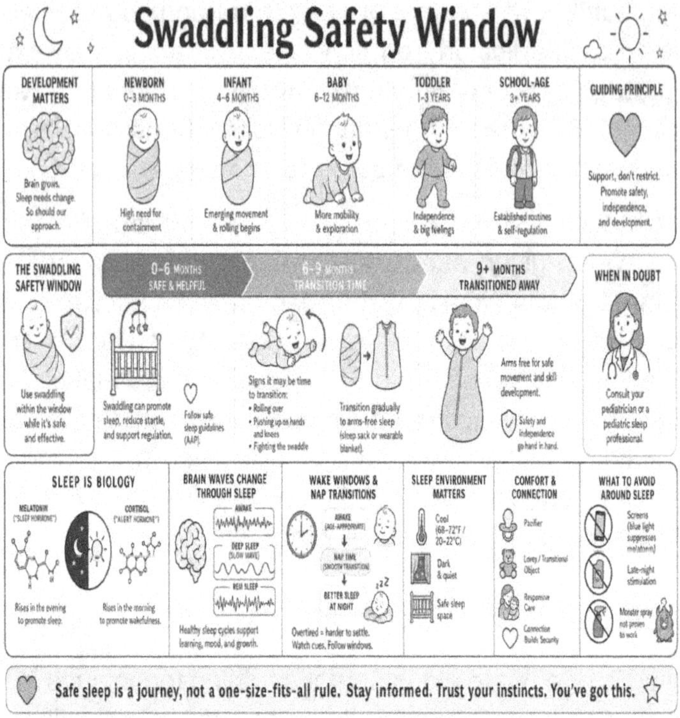

The harder conversation about swaddling is the cutoff, and it is non-negotiable. Swaddling must stop when your baby shows any sign of rolling, typically between 8 weeks (on the early end) and 4 months (on the later end), though most commonly in the 10-to-14-week range. The reason is straightforward and serious: a swaddled baby who rolls to her stomach cannot use her arms to push her head up or reposition her airway. The moment you see any sign of rolling, even an early attempt with legs and hips but without full roll completion, the swaddle stops. You can transition to a sleep sack, which keeps the arms free and the body warm, or let her sleep unswaddled with appropriate room temperature and appropriate sleep surface.

Many parents dread the swaddle transition because swaddling has been working so well. The first few nights without it are often harder.

Stick with it. Most babies adapt within three to five nights, and the short-term disruption is worth it.

Feeding, Hunger, and the Sleep Connection

At this age, sleep and feeding are so tightly coupled that you really cannot think about one without the other. Newborns wake because they are hungry. They fall back asleep because they are full and comfortable. Their stomach capacity means they need to feed every two to three hours around the clock for the first several weeks. At certain points, particularly during growth spurts and developmental periods, they will cluster feed, meaning they want to nurse or take a bottle every 45 to 60 minutes for several hours in a row.

Cluster feeding is one of the most anxiety-provoking experiences for new parents because it looks exactly like what it would mean if an adult were doing it: something is wrong. "Why does she keep wanting to eat? Is my milk not enough? Is she not getting enough from the bottle? Why won't she settle?" The answer, almost always, is that cluster feeding is normal and functional. It typically happens in the late afternoon and early evening for breastfeeding families, and it is thought to serve both a supply-building function (frequent nursing signals higher demand to the supply system) and a behavioral function (loading up on calories before the slightly longer first stretch of night sleep). Formula-fed babies also cluster feed, though sometimes less intensely.

Diagram 2.9: Room-Sharing Setup

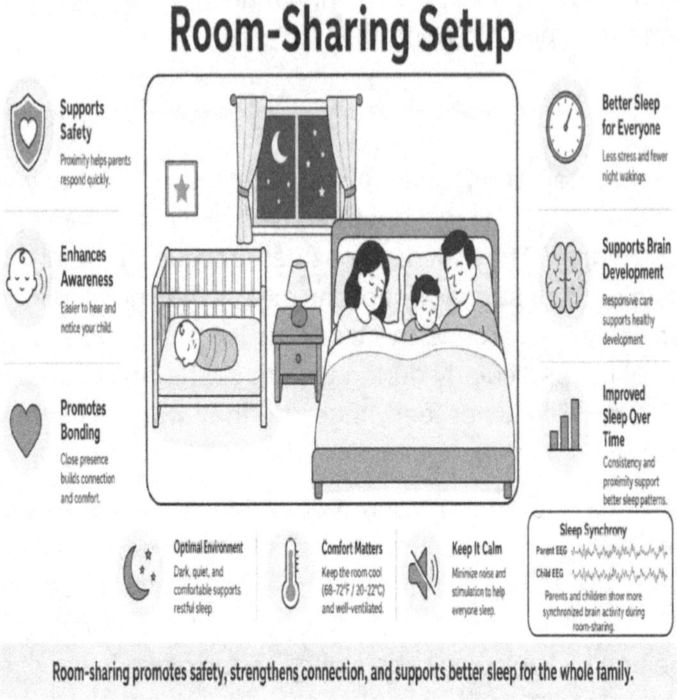

Growth spurts follow a rough but recognizable pattern: they tend to cluster around two weeks, three weeks, six weeks, three months, and six months. During these windows, the baby will feed more, sleep more (or less, depending on the baby), be harder to soothe, and generally behave in ways that suggest something has changed. Something has changed. She is growing. This is supposed to happen. It typically lasts two to five days and then settles, and after a growth spurt you will often notice that the baby seems more capable in some way, more alert, slightly stronger, occasionally showing a new skill.

Hunger cues before crying are worth learning because responding earlier, when the cue is subtle, tends to lead to a calmer feed and an easier return to sleep. Early hunger cues include rooting, which is side-to-side head turning with mouth opening; sucking on hands or fingers; increased alertness and arm movement; and fussing that is not yet a full cry. A baby who has reached a full crying hunger cue is harder to latch, harder to bottle feed effectively, and takes longer to settle. If you can catch the early cues, middle-of-the-night feeds become faster and quieter, which matters a lot when you are running on two hours of sleep.

The "Good Sleeper" Who Gets Harder

Here is something nobody warns parents about, and it causes tremendous unnecessary guilt when it happens: some newborns are remarkably easy in the first four to six weeks, sleeping long stretches, rarely crying, going down without a fuss. Their parents feel like they have done something right, like they have figured it out. And then, around six to eight weeks, things get harder. More waking. More fussing. Shorter naps. It feels like regression, like something went wrong.

What actually happened is developmental. The early easy period was partly a product of the newborn "sleepy phase," a normal and temporary state of high sleepiness and low alertness that characterizes many babies in the first two to three weeks. As the baby becomes more neurologically awake, more aware of her environment, and more susceptible to overstimulation, sleep often becomes harder before it gets easier again. The six-week mark is also when the SCN starts calibrating, meaning the brain is doing more active work in sleep-wake regulation, which can temporarily increase wakefulness.

This is normal. It is not a failure. It is not because you started a bad habit. It is neurology, working through a predictable sequence. The pattern typically improves again around 10 to 14 weeks as melatonin rhythm emerges and sleep consolidates slightly. If you are in this harder phase right now, just knowing it has a predictable end can make it feel more bearable.

Protecting Your Own Sleep

Everything in this chapter is about the baby's biology, but you are also a biological organism with sleep needs, and those needs do not disappear because there is a new person in the house. They actually become more important, because sleep deprivation at the level of new parenthood is measurable cognitive and emotional impairment. Studies on postpartum sleep deprivation consistently show that it contributes to difficulty with decision-making, emotional regulation, pain sensitivity, and, most seriously, postpartum depression and anxiety.

The most impactful single thing you can do in these first weeks is protect at least one unbroken block of sleep of three to four hours somewhere in every 24 hours. "Sleep when the baby sleeps" is the advice, and it is partly right and partly impossible, because the baby sleeps in 50-minute bursts and you cannot cycle through your own sleep architecture in 50 minutes. But if you can find a way to hand off responsibility for one longer block, even on alternating nights, the difference in cognitive function is dramatic.

For couples, the most common workable version is a split shift: one parent handles all wake-ups from, say, 9 p.m. to 2 a.m., and the other handles everything from 2 a.m. to 7 a.m. The off-duty parent sleeps in a different room with the door closed if possible, or uses earplugs if a different room is not available. This does not mean both parents get a full night; it means each parent gets one protected block. For single parents, the block may come from a trusted family member, a postpartum doula, or a scheduled daytime stretch while someone else holds the baby.

One more thing, and this one matters: postpartum depression and postpartum rage are real, common, and undertreated. If you are experiencing thoughts of harming yourself or the baby, if you feel a rage toward your child or your partner that frightens you, if you feel detached, hopeless, or like you are not bonding with your baby the way you expected, please tell your OB, midwife, or pediatrician at your next appointment. You do not need to be in crisis to ask for help. These are treatable conditions, and getting support is not a sign that you are failing as a parent. It is a sign that you are taking care of yourself so you can take care of your baby.

Try This Tonight

For this age range: 0 to 3 months

1. At the next nighttime feed, keep all lights dim or off and avoid talking or making eye contact with the baby. Use a low-level red-spectrum nightlight if you need to see. This begins teaching the difference between night feeds and daytime interactions, even before her SCN can fully register it.

2. If you are swaddling, check the hip room tonight. Slide two or three fingers between the swaddle fabric and the baby's thighs when the knees are bent up. If you cannot fit those fingers in, the hips are wrapped too tight. Adjust to allow natural frog-leg positioning at the hips.

3. Look for the early hunger cues tonight: rooting, hand-sucking, increased arm movement, soft fussing. Try responding at the early cue stage rather than waiting for a full cry and notice whether the feed is faster and easier.

4. Review the sleep space together with your co-parent or a trusted helper. Remove anything that should not be there: loose blankets, positioners, wedges, soft items of any kind. The surface should be firm, flat, and covered only by a fitted sheet made for that specific product.

5. Try the 30-second wait before responding to sounds from the baby tonight. If she stirs, watch and listen for 30 to 60 seconds before going in. Many nighttime sounds are Active Sleep movement and will resolve on their own. If she escalates to a full cry, go in immediately.

6. Designate a sleep block for the primary nighttime caregiver tonight. Even a two- to three-hour stretch of truly uninterrupted sleep, in a separate room with the door closed, will make a measurable difference. Write down who is on duty and for which hours, so there is no middle-of-the-night negotiation.

7. If you have not already, add your baby's safe sleep setup to the mental checklist you run before leaving the room at every nap and every overnight placement: back down, flat firm surface, nothing in the space, temperature comfortable, in your room.

Parent Takeaways

- Newborn sleep is biologically scattered across 24 hours because the circadian clock has not yet calibrated; this is normal until around 8 to 12 weeks and does not reflect anything you are or are not doing.
- Active Sleep looks alarming but is critical for brain development; wait 30 to 60 seconds before responding to movements and sounds to avoid interrupting a sleep cycle the baby is completing on her own.
- Sleeping through the night is not biologically achievable in the first 12 to 16 weeks; a more realistic goal is one stretch of three to four hours in the early night, and that is a genuinely good outcome.
- Every sleep placement follows the same safe sleep checklist: back down, firm flat surface, nothing in the sleep space, in your room on a separate surface, pacifier offered after breastfeeding is established.
- Swaddling must stop the moment any sign of rolling appears, regardless of how well it has been working.
- Protecting one unbroken three-to-four-hour sleep block for the primary caregiver every 24 hours is the single highest-leverage

thing you can do for your own wellbeing in this stage, and your wellbeing is part of your baby's care.

3 The Clock Starts Ticking

It started on a Tuesday, though honestly the exact day blurs when you are this tired. Your baby had been giving you five hours at night, not perfect sleep by anyone's definition, but enough. Enough to feel like a person again. Enough to have one coherent thought before noon. And then, seemingly out of nowhere, it was gone. The five-hour stretch vanished. The ninety-minute wake-ups returned with a kind of cheerful relentlessness, as if sleep had never improved at all. You lay there in the dark at 2 a.m., then 3:30, then 5, trying to remember if you had done something differently. A new swaddle. A room that was too warm. Too much stimulation before bed. You ran through the list twice and came up empty.

When you mentioned it to your pediatrician, they gave you a sympathetic nod and two words: four-month regression. And then you were back in the parking lot, buckling a screaming baby into a car seat, no more informed than when you walked in.

Here is what the pediatrician probably did not have time to tell you: this is not a regression at all. The word implies something went backward, that your child lost a skill or slid down a developmental ladder rung. What is actually happening is the opposite. Your baby's brain is reorganizing sleep at a fundamental level, building toward the sleep architecture humans maintain for the rest of their lives. The construction process is noisy. It produces more wake-ups. It feels like backsliding. But underneath the exhaustion and the 3 a.m. confusion, something remarkable is being built.

You are also, right around now, standing at the threshold of what a lot of parents call sleep training territory. The internet has opinions about this, loud ones, delivered with the confidence of people who have not met your baby or spent a single night in your house. The research, it turns out, is considerably less dramatic than the online debates suggest. Several approaches work. The differences between methods are smaller than those between consistent and inconsistent execution. And none of the evidence-supported options, when used appropriately and in the right developmental window, cause the lasting psychological harm that the most heated forum posts predict.

This chapter is going to give you the biology, the developmental timeline, the honest summary of what the research actually says, and a clear-eyed look at the choices in front of you. You get to decide. That is true now, and it will be true at every stage of this process.

Chapter Summary

This chapter covers the most significant neurological event in your infant's first year of sleep development: the maturation of sleep architecture that happens between roughly three and five months. Around this time, the suprachiasmatic nucleus, the brain's internal clock, comes properly online and begins driving a real circadian rhythm. Melatonin secretion shifts from nearly flat to a genuine day-night pattern. And sleep itself shifts from the simple two-state system of early infancy into something far more complex, with distinct light, deep, and REM stages organized into cycles that now run for approximately 90 minutes instead of the original 50.

That restructuring is what most people are calling the four-month regression. This chapter explains why the term "regression" is technically wrong, what is actually happening in brain development, and why this shift produces the wave of wake-ups that leaves so many families blindsided. It also walks through the full landscape of sleep-training approaches, from the most gradual and parent-present methods to full extinction, with honest summaries of what the evidence shows for each. The goal is not to tell you what to do. The goal is to ensure you have the information you need to make a decision that truly fits your family, temperament, and child.

What This Chapter Will Cover

- The SCN and circadian onset. Between twelve and sixteen weeks, your baby's suprachiasmatic nucleus begins producing a real circadian signal, and melatonin secretion takes on a day-night rhythm for the first time. This is the biological shift that makes a consistent sleep schedule possible, and it explains why attempts to establish one before twelve weeks often feel futile.

- Why the four-month event is a progression. At around three to four months, sleep cycles consolidate from roughly 50 minutes to 90 minutes or more, and sleep architecture reorganizes from the newborn's two-state Active/Quiet system into true REM and NREM stages. More arousals occur between cycles during this transition, which most families experience as regression.

- The shift from Active and Quiet Sleep to REM and NREM. Newborn sleep is organized around two broad states. Mature infant sleep, the kind that begins to emerge at four months, includes N1, N2, N3 deep sleep, and full REM. Understanding this transition helps you understand what your baby actually needs during night wakings.

- Drowsy-but-awake and the roots of self-soothing. The drowsy-but-awake practice is not just a tip; it is the behavioral mechanism through which infants learn to reconnect their sleep cycles independently. This section explains the neuroscience behind it and how to find the right window.

- Sleep training methods, plainly explained. Extinction, graduated extinction, the chair method, pick-up-put-down, and parent-presence fading all have published evidence behind them. This section walks through each one clearly, without advocacy or shame, so that parents can make an informed choice rather than a reflexive one.

- Night feeding in months four through six. Nutritional night feeding is real and necessary early on. Most healthy infants at adequate weight can begin consolidating night feeds by five to six months, but this is not a blanket rule. This section explains how to tell the difference between a feed that serves a nutritional purpose and one that has primarily become a sleep association.
- How to know your baby is ready. Not all four-month-olds are in the same developmental place. This section covers the weight, feeding, and developmental markers that suggest readiness for formal sleep training, along with the signals that mean waiting a few more weeks is the better call.
- Common pitfalls. Starting too early, switching methods before they have a chance to work, and parent disagreement mid-process are the three most common reasons families feel like sleep training failed. This section addresses all three directly.

Diagram 3.1: The SCN Comes Online: Circadian Onset at 12-16 Weeks

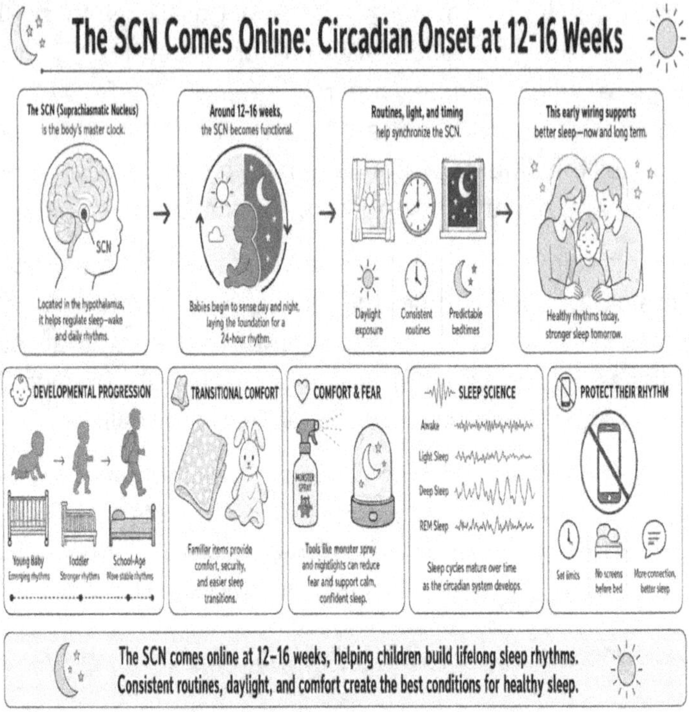

Between twelve and sixteen weeks, a small region of the hypothalamus called the suprachiasmatic nucleus, or SCN, begins functioning as a true internal clock. The SCN sits just above the optic chiasm, where the optic nerves cross, and it receives direct input from specialized retinal cells sensitive to blue light. In a newborn, this connection exists anatomically, but the SCN is not yet producing a reliable rhythmic output. Sleep happens when it happens, driven more by hunger and gut signals than by any internal clock.

By twelve weeks in most healthy infants, the SCN begins firing in a recognizable twenty-four-hour pattern. Cortisol, which promotes wakefulness, peaks in the morning. Melatonin, which prepares the brain for sleep, begins rising in the early evening and peaks in the late night. This is the physiological event that transforms your baby from a creature who sleeps in short bursts scattered across the day and night into an organism that can, at least in principle, consolidate sleep into longer nighttime blocks. The shift does not happen all at once. For several weeks, the rhythm is present but shallow, easily disrupted by irregular light exposure, late-night feedings in bright conditions, or overstimulation close to bedtime.

The New Clock and What It Needs

Because the SCN is now taking its cues from light, the environment you create around your baby in this window shapes how quickly and strongly the circadian rhythm consolidates. Morning light exposure is particularly powerful. A walk outside in the first hour after waking, or time near a bright window, sends a strong "day has begun" signal to the SCN. That signal sets the anchor point of the whole twenty-four-hour rhythm. Evening light, especially blue-spectrum light from screens and bright overhead fixtures, delays the rise in melatonin and can push natural sleepiness back by 30 to 60 minutes in infants, just as it does in adults.

None of this requires an elaborate routine. It requires a general principle: make daytime bright, social, and stimulating, and make the two hours before bedtime consistently dim, quiet, and calm. The contrast is what trains the clock. A consistent bedtime does the same work from the behavioral side. When the same sequence of events reliably precedes sleep, the SCN learns to anticipate the sleep window and begins ramping up melatonin slightly earlier each night. By six months, many families are surprised to find that their baby begins rubbing their eyes at almost the same time each evening. That predictability is not coincidence; it is the circadian system doing exactly what it was designed to do.

AAP guidance through this age band continues to emphasize the same safe-sleep environment that applied in the newborn period: a firm, flat surface free of soft bedding, bumpers, and loose items; back to sleep for every sleep; and a separate-but-close sleeping arrangement. Room-sharing, with your baby in their own bassinet or crib within arm's reach, is associated with lower risk and remains the AAP recommendation through at least six months. Bed-sharing carries separate risk considerations and is not consistent with current AAP guidance, regardless of how well-intentioned the arrangement is. This is worth restating here because the four-to-six-month window is exactly when many exhausted families consider moving the baby into their bed to get through the night. The intention is completely understandable. The risk profile does not change with intention, though, and it is worth knowing that before the exhaustion decides for you.

Diagram 3.2: 50 to 90-Minute Cycle Consolidation

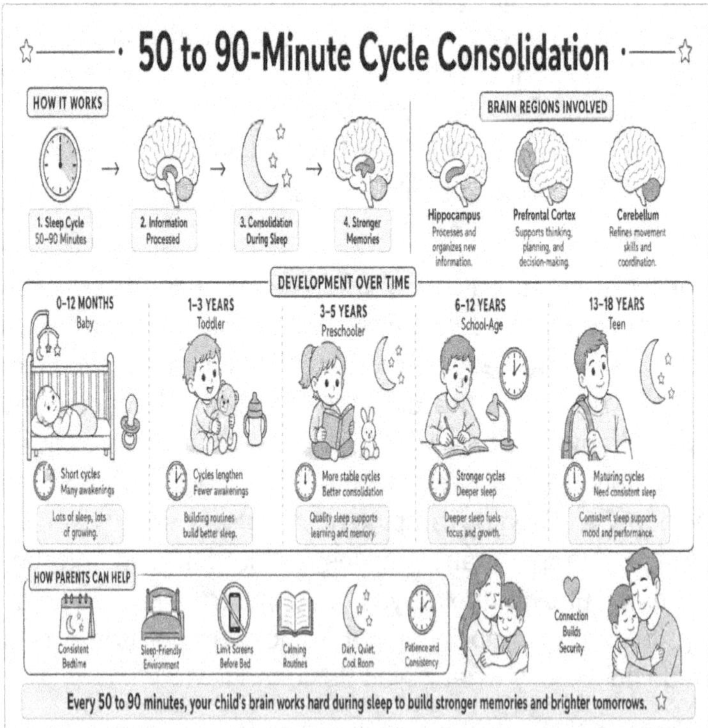

Here is the core of what is actually happening during the so-called four-month regression. A newborn's sleep cycle runs approximately forty to fifty minutes. At the end of each cycle, the brain surfaces briefly to a lighter state before descending into the next cycle. In newborns, these brief surfacings are usually quiet enough that parents sleep through them, or the baby resettles quickly on their own. The cycles are short, the transitions are shallow, and the whole system is, in some ways, forgiving.

Around three to four months, the sleep cycle lengthens toward the adult ninety-minute duration. At the same time, the architecture within each cycle becomes more complex. What was once a simple alternation between Active and Quiet Sleep is now beginning to differentiate into the layered stages of adult sleep: light NREM (what adults know as N1 and N2), deep slow-wave sleep (N3), and full REM. That architectural complexity is neurologically sophisticated and developmentally excellent. The timing, though, is rough. The longer cycles and more articulated stage transitions produce partial arousals that are more conscious than those produced by the newborn brain. Babies who were resettling quietly through cycle transitions at two months may now cry out, look around, and genuinely wake at every transition, several times a night.

The Four-Month Progression

Calling this event a regression is understandable from a parent's perspective. From a neuroscience perspective, it is precisely backward. The four-month shift is the brain maturing. Sleep is not getting worse at a fundamental level; it is getting more adult. The brief disruption in consolidated sleep happens because the new architecture is not yet fully integrated with the self-soothing skills your baby needs to navigate cycle transitions smoothly. The brain is ahead of the behavior, essentially.

This matters practically because it tells you what the actual goal is. You are not trying to get back to the four-month-ago sleep schedule for your baby. You are helping your baby's developing brain learn to use the new architecture smoothly. That means the four-month period is not a problem to fix by waiting it out. It is a developmental opening, a moment when the newly maturing sleep system is actually primed to learn new associations. The skills your baby builds during this window, specifically the ability to connect sleep cycles without a parent's physical intervention, will serve them for years.

Diagram 3.3: The Four-Month Progression, Not Regression

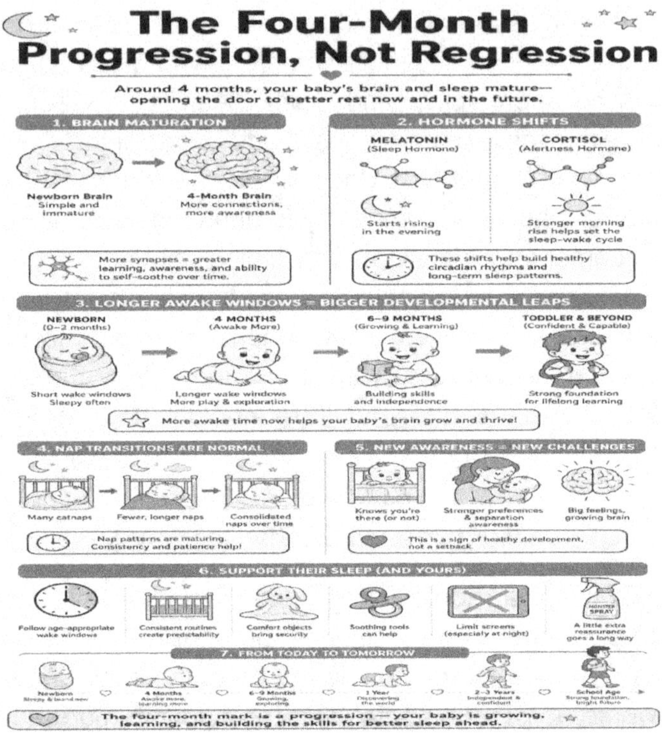

The distinction between Active Sleep and Quiet Sleep in newborns and the emerging REM/NREM distinction in older infants is worth understanding clearly, because it explains a lot of the behavior you see at night. In Active Sleep (the newborn precursor to REM), the baby's eyes flutter under closed lids, limbs twitch, facial expressions cycle through what appears to be a full range of emotions, and breathing is irregular. Many parents pick up a baby in Active Sleep, thinking they are waking, only to have the baby cry vigorously because external stimulation pulls them out of a sleep state they would have navigated independently.

Quiet Sleep in newborns is the precursor to NREM. Breathing is regular, the body is still, and the brain is doing the consolidation work that deep sleep does in adults. By four months, these two broad states begin to differentiate into more granular stages. N2, the light-to-medium NREM stage that adults spend the most time in, begins appearing as a distinct electroencephalographic signature. Slow-wave N3 sleep deepens and becomes more reliable. Full REM, with its characteristic rapid eye movements and near-complete muscle paralysis, becomes a consistent part of the sleep architecture.

Diagram 3.4: REM-NREM Maturation at 4 Months

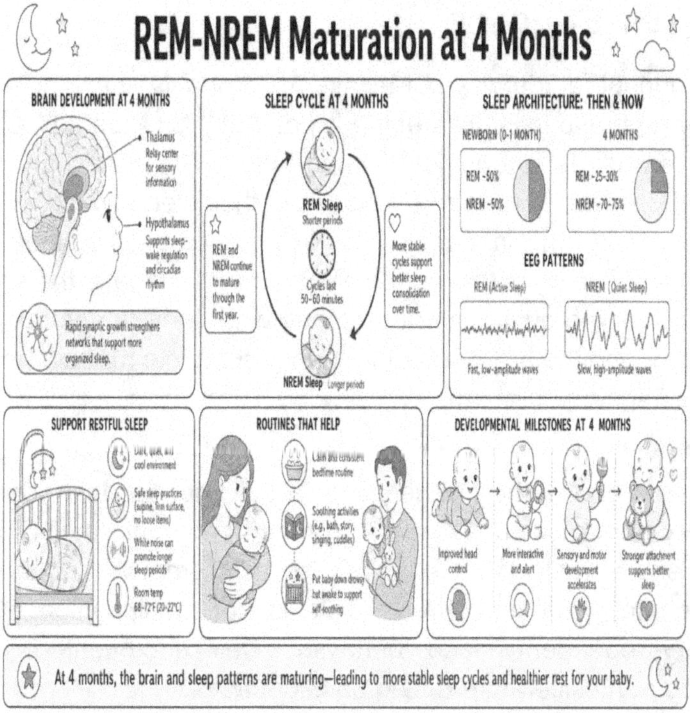

What this means for parents is that the range of behaviors you see during sleep widens. A four-month-old in N3 slow-wave sleep can be remarkably hard to rouse and will sleep through the sound of a passing truck. The same baby forty minutes later, surfacing from a REM period, may cry out, look around with open and apparently alert eyes, and seem fully awake for a minute or two before the next cycle begins. If you respond immediately and fully to that transition wake, feeding or rocking the baby back to sleep, you are linking the cycle transition to your presence. The baby learns, reasonably, that the end of a sleep cycle means you will appear. That association is not a moral failure. It is a completely normal learned behavior, and it is exactly what the drowsy-but-awake approach is designed to prevent or gently undo.

Drowsy-But-Awake and the Formation of Self-Soothing

The concept of drowsy-but-awake is probably the single most repeated phrase in infant sleep advice, and it is repeated that often because it is genuinely important. The mechanism is straightforward: if your baby is placed in their sleep space while still slightly awake, the last conscious experience before sleep is the sensation of being in that space. When they surface at the end of a sleep cycle, the environment matches what they remember falling asleep in, and the brain can more easily complete the reconnection without flagging an alarm.

If, on the other hand, your baby falls fully asleep in your arms or at the breast and is then transferred to the crib, the last conscious memory before sleep is warmth, contact, and your smell. The crib at the end of a sleep cycle is therefore a mismatch, and the brain appropriately signals distress. This is not manipulation. This is memory-dependent sleep behavior, well documented in the developmental literature dating back decades.

Diagram 3.5: Drowsy-But-Awake Window

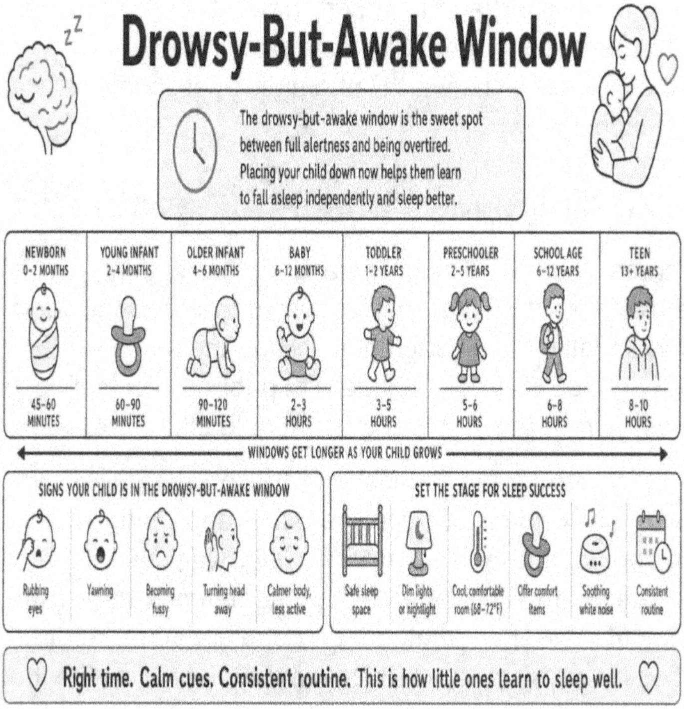

Finding the drowsy-but-awake window takes practice, and missing it in either direction is common and fine. If you put your baby down too awake, they may cry with full energy and be unable to settle. If you put them down too asleep, you lose the benefit. The window you are looking for sits between those two states: eyes heavy, possibly drooping; limbs loose; brief flutter of the eyes; a slowing in sucking if they have a pacifier. The baby is aware of the transition from your arms to the crib but not alarmed by it. It may take ten or twenty tries to consistently find this window. That is normal.

The drowsy-but-awake practice matters even if you have no current plans to do formal sleep training. Every time your baby completes a sleep-cycle transition independently, even a brief one, they build neural pathways for self-soothing. Those pathways strengthen with repetition. A baby who sometimes falls asleep independently, even if inconsistently, is further along this developmental track than a baby who has never had the opportunity. Which means that even if full sleep training is not right for your family right now, the drowsy-but-awake practice is worth attempting at bedtime whenever circumstances allow.

Sleep Training Methods: A Clear Map

The phrase "sleep training" means different things to different people, and much of the conflict swirling around the topic stems from people talking past each other using the same words. Here is a plain-language map of methods with published evidence, ordered from most parent-present and gradual to least.

Parent-presence fading begins with the parent fully in the room, often sitting next to or touching the baby, and gradually moves the point of contact further away over days or weeks until the parent is outside the room. This approach is sometimes called the Sleep Lady Shuffle or chair method in popular literature. It requires patience and consistent execution over a longer timeline than more abrupt approaches, but it keeps the parent emotionally present throughout the process. Many infants and parents with higher sensitivity who find separation acutely distressing respond well to this method. The primary challenge is consistency: fading too slowly, or returning to earlier stages when the baby protests, can extend the timeline considerably.

Pick-up-put-down is exactly what the name suggests. The parent puts the baby in the crib drowsy-but-awake, and if the baby cries beyond a certain threshold, picks them up to calm, then puts them down again once calm, before sleep returns. This approach is associated with Kim West's work and various iterations in parent-focused sleep resources. It works well for some infants, particularly in the four- to five-month range, when cycles are just beginning to consolidate. It is more physically demanding than other methods because the number of pick-ups in the early nights can be high, and some infants become more stimulated by the pick-up itself, making calm-down harder. It is not a faster method than others, but it gives parents something concrete to do.

Diagram 3.6: Gentle Sleep Training Methods Map

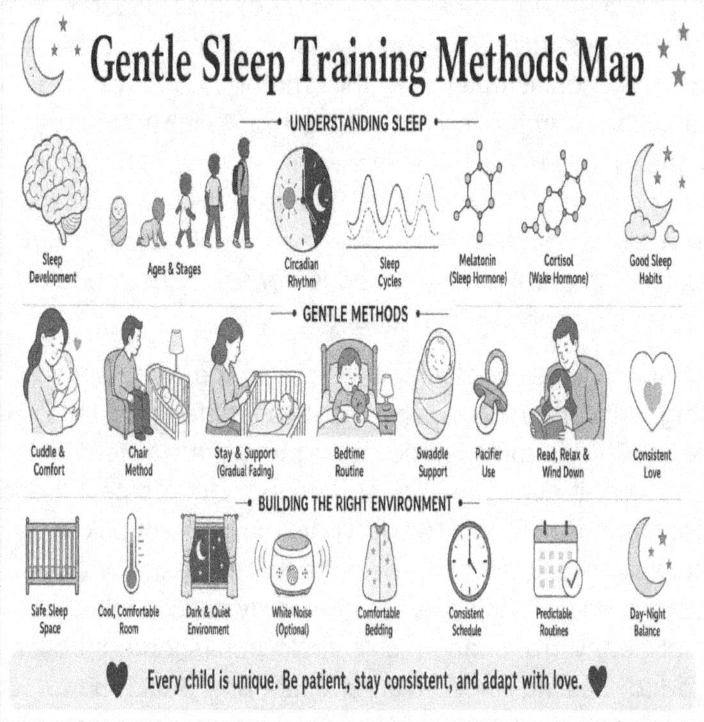

The Pantley pull-off, from Elizabeth Pantley's No-Cry Sleep Solution, specifically targets the feeding-to-sleep association. Rather than breaking the association abruptly, it involves gradually reducing the amount of feeding that precedes sleep, interrupting the suck-to-sleep pattern before the baby is fully asleep, and repeating as needed to allow the baby to complete the transition without the breast or bottle. This approach is slower than any method that involves crying and requires a great deal of parental patience and attention to detail during execution. It works. The timeline is longer. For families in which any amount of crying is genuinely unmanageable, it is a legitimate and evidence-based option.

Graduated extinction, most commonly associated with Richard Ferber, though the research base extends well beyond any one author, involves putting the baby down drowsy-but-awake and waiting progressively longer intervals before going in for a brief check. The check-in is brief and non-feeding, intended to reassure without fully resettling. The intervals lengthen night by night. In families that consistently use this method, behavioral research generally shows improvement within 5 to 7 nights for most infants within the appropriate developmental window. The initial nights involve crying that many parents find very difficult to tolerate. The literature also documents elevated parental cortisol during this process, which means the method is genuinely hard on caregivers even when it is working as intended.

Full extinction involves placing the baby in the sleep space and not returning until morning or the established feeding time, without check-ins. This is the method most people are referring to when they say "cry-it-out." It tends to produce the fastest behavioral change, often within three to five nights, because check-in visits can function as intermittent reinforcement, thereby extending the crying. For parents who can tolerate the genuinely difficult initial nights, the speed of behavioral change is real. No published controlled study has found lasting negative effects on attachment, cortisol regulation, or emotional development in healthy infants when this method is used at the appropriate developmental age, and multiple randomized controlled trials and long-term follow-up studies now confirm this finding.

Diagram 3.7: Ferber vs. Full Extinction Comparison

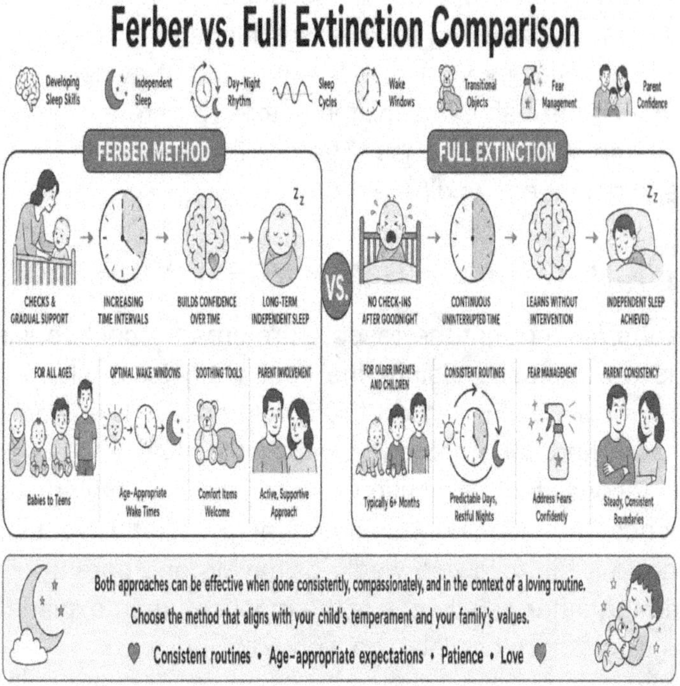

Let's be direct about the attachment question because it carries the most emotional weight. The fear is that allowing a baby to cry without immediate response will cause lasting psychological harm, damaged attachment security, or elevated stress hormones. This is a reasonable concern and one the research has taken seriously. Multiple studies have followed children whose parents used graduated or full extinction methods in infancy and compared them with control groups on measures of attachment security, behavioral adjustment, cortisol levels, and parent-child relationship quality at ages 1, 2, 3, and 5. The consistent finding across these studies is that no significant differences emerge between groups on any of these measures. Attachment security is built over thousands of interactions across months and years. It is not undone by several nights of crying in a safe sleep environment.

This does not mean the methods feel good. Many parents describe the early nights as among the hardest things they have had to do. Feeling distressed by your baby's crying is not evidence that you are harming your baby. It is evidence that you are a normal parent with a normal stress response. The two things are different, and it matters to hold them as different.

The Right Method Is the One You Will Use

Here is the finding that the research returns to most consistently, and it is worth stating plainly: the best predictor of sleep training success is not which method the family chose. It is whether the family executed the method consistently over a sufficient trial period. Inconsistent execution, specifically switching between approaches, abandoning a method at the first hard night and returning to it the following week, or having one parent follow the plan and the other quietly not, is the most common explanation for families who report that "nothing worked."

Two weeks of consistent execution is the minimum meaningful trial for most methods. Three weeks is a more realistic window for gradual approaches. If you are switching strategies every three or four days because the current one is not delivering instant results, you are not giving any method a chance to work. Behavioral change in an infant follows a reliable curve: things often get harder before they get easier, a phenomenon the behavioral literature calls an extinction burst. If you abandon a method at exactly the point where the extinction burst is most intense, you have put your baby through the hard part with none of the benefit.

This is not a criticism. It is enormously difficult to listen to your baby cry and hold a line, especially at 2 a.m. after months of interrupted sleep. The more prepared you are for the extinction burst before it happens, the more likely you are to hold course through it.

It also helps to have a clear definition of success before you begin, rather than letting the goalposts shift each night. "Success" does not mean zero wake-ups. For most families starting in the four-to-six-month window, it means a baby who can return to sleep independently after a cycle transition without requiring a parent's physical intervention every time. It may still mean one nutritional feed. It may still mean occasional nights where illness or a leap disrupts the pattern. Defining success as "independent settling at the beginning of the night and at cycle transitions" gives you something concrete to measure and helps prevent the frustration of holding the plan against an unrealistic standard.

Night Feeding in Months Four Through Six

Night feeding is not a sleep training problem. It is a nutritional reality, and it is important to separate the two. In the first four months, most babies genuinely need one to three night feeds depending on birth weight, feeding method, and individual growth trajectory. Treating these feeds as a "bad habit" to be eliminated is not appropriate in that window. They serve a biological function.

By five to six months, the picture changes for most healthy infants. Total daily caloric intake for a typically developing baby at this stage can generally be met within daytime hours. The gastrointestinal system has matured enough to process larger feed volumes, stomach capacity has grown, and the caloric density of breast milk or formula is sufficient to support a longer nighttime fast. At this point, some night feeds shift from nutritional necessity to learned association. The baby is not waking because they are hungry; they are waking because they have learned that waking is the cue for a feed, and the feed is soothing.

Diagram 3.8: Night-Feed Phase-Out Plan

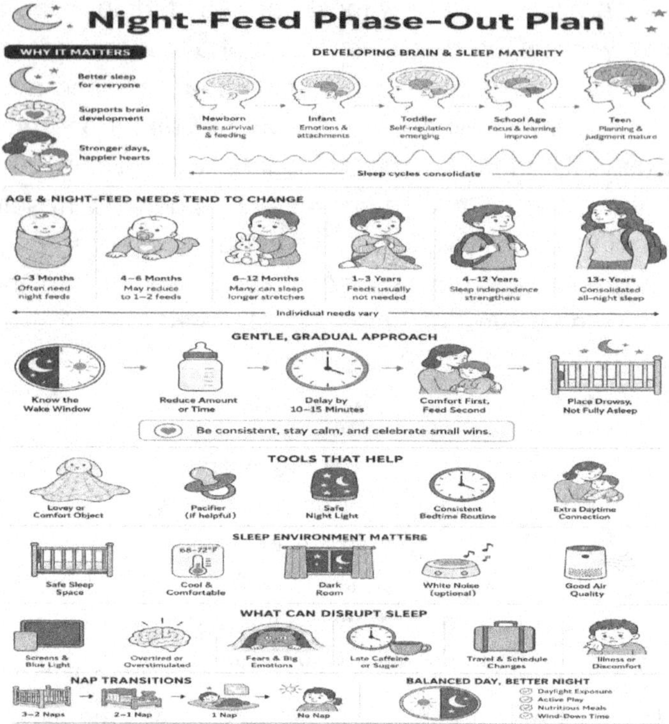

Distinguishing between a nutritional feed and a comfort feed is not always easy, and your pediatrician's input here is genuinely valuable. A few practical markers help. A baby who takes a full feed at 2 a.m. and then refuses the morning feed is likely getting meaningful calories at night. A baby who nurses for four minutes, falls immediately back to sleep, and wakes again two hours later is more likely to be using the feed as a settling mechanism than to meet a genuine caloric need. Birth weight history, growth curves, and your pediatrician's overall assessment should all factor into this determination. No book can tell you whether your specific baby, with their specific weight and growth history, has crossed the threshold of nutritional necessity. Your pediatrician can.

When the decision is made to begin reducing night feeds, gradual reduction is gentler than abrupt elimination. Decreasing feed duration by one to two minutes per night, or decreasing formula volume by half an ounce every two nights, allows the body to adjust and reduces the physiological hunger signal that would otherwise produce genuine distress. Abrupt elimination of a feed a baby still depends on metabolically will produce crying that is not the sleep-training kind; it is hunger, and it needs to be met.

How to Know Your Baby Is Ready

The question of developmental readiness is the one that prevents the most harm in this whole topic. Sleep training before a baby's circadian and sleep architecture systems are sufficiently mature produces poor results and unnecessary distress for no developmental benefit. The general guidance from pediatric sleep researchers places the lower bound for any formal behavioral sleep intervention at four months corrected age, with five to six months being a more conservative and widely recommended starting point.

The markers of readiness are not a checklist to run through in one visit. They are a composite picture. Your baby should be back to or above birth weight and growing consistently on their growth curve. Night feeds should have been addressed with your pediatrician, and any reduction plan should be in place or already underway. Your baby should already be showing some capacity for self-soothing, however brief. You should have pediatric clearance for any planned feeding changes. Your baby should be healthy, without active illness, recent vaccinations in the last forty-eight hours, or other sources of physical discomfort that would confound any behavioral intervention.

Diagram 3.9: Choosing the Right Method for Your Family

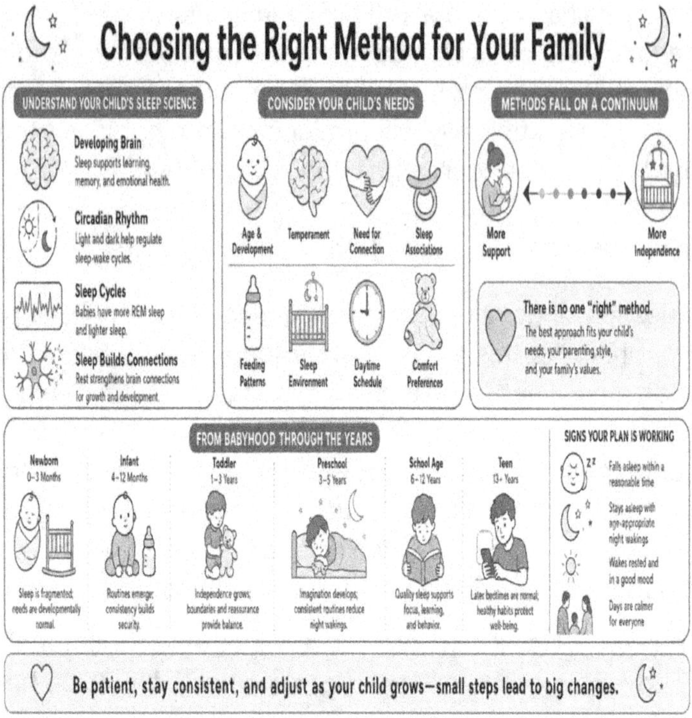

sz

Temperament matters here as much as any clinical criterion. By four months, most parents have a clear enough read on their baby's temperament to factor it into method selection. A high-sensitivity infant, one who escalates quickly, de-escalates slowly, and seems to need more physical contact to regulate, will likely have a longer and harder experience with full extinction and a better one with gradual fading or parent-presence approaches. An easier-temperament infant who calms relatively quickly and accepts comfort from a wider range of sources may do well with almost any consistent method. Neither temperament type makes the baby better or worse. They make certain methods a better fit.

Common Pitfalls in This Window

Starting too early is the most consequential mistake. Before twelve weeks corrected age, the circadian system is not operational, and before four to five months, sleep architecture maturation is not complete. Behavioral interventions in this window do not produce the changes parents hope for because the underlying biology has not yet supported them. If you are at two or three months and considering sleep training because you have read that it works, the evidence is asking you to wait. The waiting is hard. It also protects your baby from an experience that serves no developmental purpose at this stage.

Inconsistent execution, as discussed above, is the second most common pitfall. This includes what might be called split-household execution, in which one caregiver follows the plan, and the other does not. The baby does not experience this as "one parent follows the rules, and one does not." The baby experiences the plan as unpredictable and, therefore, learns to escalate protest because escalation sometimes works. If one partner is not on board with a method, it is far better to choose a different method that both parents can commit to than to proceed with a method one parent will undermine under pressure.

Developmental timing mismatch is the third most common pitfall and can occur in both directions. Starting too early has been covered. Starting during a developmental leap, a period of illness, a major transition like a move or a new caregiver, or right before or after a vaccination is also poor timing. Sleep training relies on behavioral learning, and learning is disrupted by stress, illness, and developmental reorganization. Giving your baby a stable two-week window, free of major disruptions, significantly improves the likelihood that a chosen method will work as expected.

A fourth pitfall that deserves naming is comparing your baby's trajectory to another family's. Social media is extraordinarily effective at surfacing the outlier: the baby who slept through the night at ten weeks, the family whose five-night Ferber run produced twelve-hour stretches, the gentle parent whose no-cry approach worked in two weeks flat. These stories are real and not typical. The typical experience is messier, slower, and more iterative. If you are consistently executing a method and seeing gradual improvement over two to three weeks, you are succeeding, even if another family's baby is ahead on a schedule you found online. Your baby has their own developmental timeline, and the job is to support that timeline, not race it.

Try This Tonight

For this age range: 3 to 6 months

1. Tonight at bedtime, observe the drowsy-but-awake window. When your baby is nursing or being held to sleep, watch for the moment when sucking slows, limbs go loose, and eyes begin to droop. That is your target window. Try placing them in the crib at that moment rather than waiting until they are fully asleep.

2. Before the next night feeding, spend sixty seconds deciding whether the feed is likely nutritional or comfort-based. You do not need to change anything tonight. Just begin paying attention to the length of the feed, your baby's response at morning feeding, and the gap between night wake-ups.

3. Write the name of the method you are currently using, or considering using, on a piece of paper and put it on the refrigerator. Make it visible. This is the beginning of the two-week commitment.

4. Dim every light in your home to the lowest usable level for the ninety minutes before your baby's bedtime. If you have overhead lights on at full brightness, switch to lamps. If you have screens on, reduce their brightness. This one environmental change begins signaling the SCN to produce melatonin earlier in the evening.

5. Check that your baby's sleep space still meets current AAP safe-sleep guidelines. Firm, flat surface. No soft bedding, pillows, or bumpers. Back to sleep. Room-sharing if possible. This is always worth a brief weekly check at this age.

6. If you have a co-parent or other nighttime caregiver, have a ten-minute conversation tonight about which method you are using and what consistency looks like. Agree on what each of you will do if the baby cries at 2 a.m. Write down the agreement if that helps.

7. Permit yourself to find this hard. The four-month window is developmentally intense, and the sleep deprivation that precedes it has been accumulating for months. Choosing a path forward and taking the first step tonight, even a small one, matters.

Parent Takeaways

- The four-month "regression" is a genuine developmental progression; sleep architecture is maturing toward adult organization, and the disruption is a sign of healthy neurological development.

- The SCN comes online between twelve and sixteen weeks, making this the right window to begin building consistent light cues and a predictable bedtime routine.

- Multiple sleep training methods have solid evidence behind them; differences in outcome are driven more by consistency of execution than by which method a family chooses.

- Evidence-based sleep training, used in the appropriate developmental window, does not damage attachment, increase long-term cortisol, or harm emotional development in healthy infants.

- The best sleep training method is the one both caregivers will stick with for a minimum of two weeks; consistency matters more than the specific approach.

- Night feeds serve genuine nutritional purposes through early infancy; the shift to comfort-based waking typically happens around five to six months, and any feeding changes should involve pediatric guidance.

4 Naps, Schedules, and the First Year of Real Routine

Your nine-month-old napped beautifully at ten and at two last week. This week she wants nothing to do with the two o'clock nap. By six in the evening she is crawling around the living room with the wild-eyed energy of someone who has had far too much coffee, and then she completely falls apart at seven and cannot settle for another forty-five minutes. You are standing in the kitchen, wondering what you did wrong, what changed, and whether what used to work will ever work again.

Nothing is wrong. She did not forget how to sleep. She did not decide, in some tiny deliberate act of rebellion, to ruin your afternoon. What you are watching is one of the most predictable biological transitions of the first year happening in real time, and the only reason it feels like chaos is that nobody warned you it would look exactly like this.

Nap biology is not a parenting philosophy. It is not a reflection of how consistent or inconsistent you have been. It is a set of neurological and hormonal processes that unfold on a developmental schedule. That schedule does not care about your work calendar, your toddler's daycare schedule, or the fact that you were finally, finally getting good afternoons. Transitions happen when your child's brain is ready for them, and the most useful thing you can do is recognize them, understand what is driving them, and know what to do next.

This chapter is for the parents of infants and toddlers who are trying to build a workable day. It covers the biology of daytime sleep, why naps are not optional at certain ages, how wake windows work as a planning tool, and how to navigate each of the major nap transitions without losing your mind. It also talks honestly about the imperfect reality of daycare schedules, traveling with a baby, and the stretches of developmental chaos that temporarily make everything harder.

You are going to learn something in this chapter that genuinely helps. That is a promise. And it starts with the science.

Chapter Summary

This chapter covers nap consolidation from early infancy through the toddler years, tracing the predictable path from many short naps in the newborn period through the 3-nap day, the 2-nap day, the 1-nap day, and eventually the full drop of daytime sleep around ages three to five. It explains the biology behind why daytime sleep is neurologically required at certain ages, and why removing naps too early produces the paradoxical result of a child who is both overtired and harder to get to sleep.

Parents will come away with a concrete framework for structuring the day: wake windows by age, the role of a consistent morning anchor, the relationship between nap timing and night sleep quality, and the signals that tell you a transition is real versus a temporary hiccup caused by illness, teething, or a developmental leap. The chapter also addresses the working-parent reality of daycare nap schedules and offers practical language to advocate for your child's needs in a shared-care environment.

What This Chapter Will Cover

- Wake windows as a daily planning tool. The amount of time a child can comfortably stay awake between sleep periods is one of the most reliable guides you have in the first eighteen months of life. Wake windows are shorter in early infancy and lengthen gradually as the brain matures. Understanding them takes a lot of the guesswork out of figuring out when to put your child down.
- The biology of homeostatic sleep pressure. Daytime sleep is not just about tiredness in a vague sense. It is driven by a specific neurochemical called adenosine that accumulates in

the brain during waking hours and creates pressure to sleep. Understanding this process explains why skipping naps leads to overtiredness rather than better night sleep.

• The 3-to-2 nap transition at around six to nine months. This is often the messiest transition of the first year, because the timing is inconsistent across children and the signs can look like plain old sleep resistance. This chapter walks through the signals and the strategy.

• The 2-to-1 nap transition at around fourteen to eighteen months. One of the rockier transitions, because the single nap is often too short for a while and compensatory early bedtimes are needed. Knowing the pattern prevents the mistake of forcing the transition before the child is ready.

• The 1-to-0 nap transition at around three to five years. This transition is the most variable of all and is often driven partly by social and childcare factors. The chapter explains the signs of readiness and what quiet rest time looks like when naps are no longer available.

• Schedule-based, wake-window-based, and cue-based approaches. These are the three main frameworks parents use to organize the nap day, and each has genuine strengths. The chapter lays them out clearly so you can choose the one that fits your family, or blend elements of more than ono.

• Catnaps, false starts, and the 45-minute intruder. Short naps and early morning wakings are among the most common complaints in early infancy. Understanding why they happen makes them less alarming and points toward practical solutions.

The Biology of Daytime Sleep

Before we talk about schedules and transitions, it helps to understand why naps exist in the first place, because the answer is more interesting than most people expect.

The brain has a mechanism for tracking how long it has been awake. Every hour of wakefulness, neurons in the brain accumulate a metabolic byproduct called adenosine. As adenosine levels rise, a signal builds throughout the nervous system that scientists call homeostatic sleep pressure, or Process S. When that pressure becomes high enough, the body begins pushing hard for sleep: attention narrows, mood destabilizes, motor coordination wobbles, and body temperature starts to drop in preparation for sleep onset.

In adults, Process S builds gradually over a sixteen-hour waking day and is largely discharged by a single overnight sleep period. In infants and toddlers, the story is very different. Their brains accumulate adenosine more quickly, their adenosine-clearing systems are less efficient, and their sleep architecture produces more fragmented clearance across multiple sleep bouts. The result is a neurological requirement for daytime sleep that is not optional, not a habit, and not something you can train away without consequences.

Diagram 4.1: The Homeostatic Sleep Pressure Curve

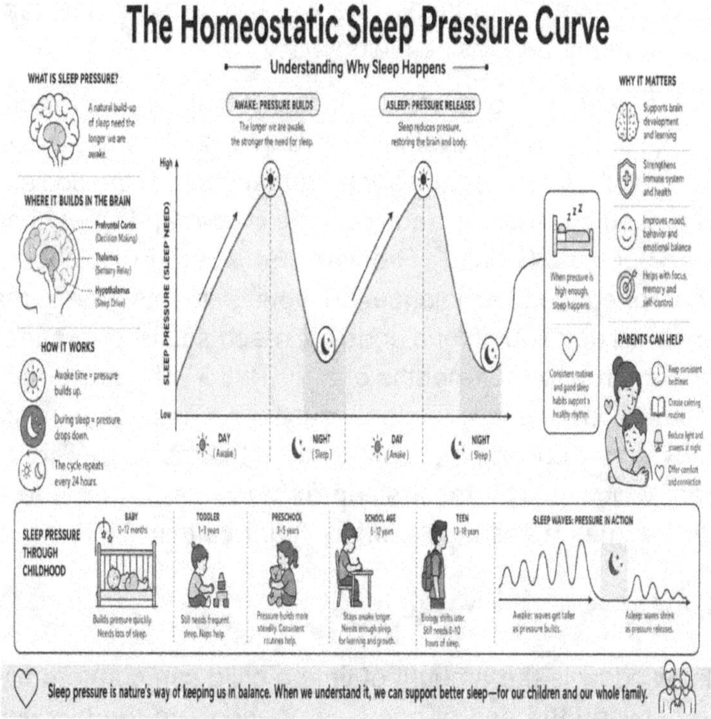

This is worth sitting with for a moment because it directly pushes back against some of the most common advice new parents receive. The idea that naps are a crutch, or that dropping them earlier will produce better night sleep, misunderstands what naps are doing biologically. A well timed nap does not steal sleep from the night. It clears adenosine buildup that, if left unresolved, would produce a cortisol spike from the stress of being overtired, and that cortisol spike is what actually ruins night sleep. The counterintuitive reality is that good naps protect night sleep in the first two years of life.

The adenosine system also explains why timing matters so much. A nap taken too late in the afternoon clears adenosine as night sleep approaches, leaving a child not tired enough at bedtime. A nap taken too early, before sufficient adenosine has built up, is either refused outright or produces a fragmented, short sleep. The window in which a nap lands, the wake window, is the practical expression of the adenosine biology.

There is a second biological clock running alongside Process S. Process C, the circadian system, is a roughly 24-hour rhythm driven by the brain's response to light and anchored by social cues including feeding, activity, and consistent wake and sleep times. In adults, Process C and Process S work in a beautifully choreographed counterbalance. In young infants, the circadian system is still developing and does not reach adult-like organization until about three to four months of age. This is why newborn sleep feels so random. It is not random exactly, but it is not yet circadian either. Nap scheduling becomes more tractable as Process C matures, which is why most sleep frameworks do not attempt to establish a true nap schedule before four months.

Wake Windows by Age

A wake window is the amount of time a child can comfortably stay awake between the end of one sleep period and the beginning of the next. It is not a hard cutoff but a biological sweet spot. Too short and the child is not tired enough to fall asleep easily; too long and adenosine accumulation tips into overtiredness, which triggers the cortisol response that makes sleep harder rather than easier.

Wake windows lengthen gradually as the brain and adenosine system mature. In the first few weeks of life, a newborn may only manage forty-five minutes to an hour of waking before needing to return to sleep. By four to six months, most infants can manage about ninety minutes to two hours between naps. By seven to nine months, windows often stretch to two and a half to three hours. Between nine and twelve months, many infants can handle three to three and a half hours, sometimes more for the stretch before bedtime.

Diagram 4.2: Wake Windows by Age

The toddler months extend wake windows further still. Around 14 to 18 months, most children move to a single nap per day and can sustain wake windows of 4 to 5 hours or longer before and after that nap. By age two, wake windows of five to six hours are typical for children still napping. Once naps are fully dropped, the wake window from morning to bedtime becomes the entire waking day, and bedtime timing becomes the main lever.

One thing that trips parents up is treating wake windows as a clock to watch rather than a child. The numbers above are averages across large developmental samples. Your child may have consistently shorter or longer windows, and either is fine as long as the sleep that follows is restorative. The goal is to use wake-window ranges as a starting framework and then calibrate from there based on what you see: how your child falls asleep, how long they sleep, and how they behave in the hour before the next nap is due. A child who is rubbing their eyes and slowing down at the ninety-minute mark is telling you their window is shorter than average. A child who is still bright and chatty at two hours and fifteen minutes is telling you theirs is longer.

Morning wake time functions as the anchor for the whole day. Whatever time your child wakes in the morning, that is your starting point for calculating the first wake window. A consistent morning wake time, even on weekends, is one of the most powerful stabilizers of the infant and toddler circadian system. It trains the brain's internal clock to expect sleep and wakefulness at predictable times, which makes nap onset easier, bedtime smoother, and overnight sleep more consolidated. This is the same anchor principle that underpins adult sleep hygiene; it just applies with even more force in a developing nervous system.

Why Daytime Sleep Is Not Optional

It is worth being very direct about this, because the pressure to drop naps early comes from many directions. Well-meaning relatives suggest it. Other parents report that their child dropped all naps at twelve months and sleeps perfectly. Daycare providers sometimes have group schedules that do not align with a child's individual needs. And there are days when the nap fight is so exhausting that just letting it go sounds like a relief.

Here is the neuroscience: the developing brain uses sleep, including daytime sleep, for processes that cannot be completed during night sleep alone. Synaptic consolidation, the process by which the brain strengthens new neural connections formed during learning, occurs predominantly during sleep and relies on sleep architecture that includes both light sleep and REM sleep. Memory processing, emotional regulation circuitry development, and motor skill consolidation all depend on sleep quantity and quality that, for children under age three, includes daytime sleep as a functional component.

When a child is consistently undertired during the day and overtired by evening, the resulting cortisol and adrenaline response creates a physiological state of arousal that is genuinely incompatible with easy sleep onset. Parents often describe this as a child who is "wired but tired," and that description is accurate. The stress hormones are doing their job of keeping the child alert and reactive because the brain interprets prolonged overtiredness as a stress state that requires arousal rather than rest.

Diagram 4.3: The 3-to-2 Nap Transition Map

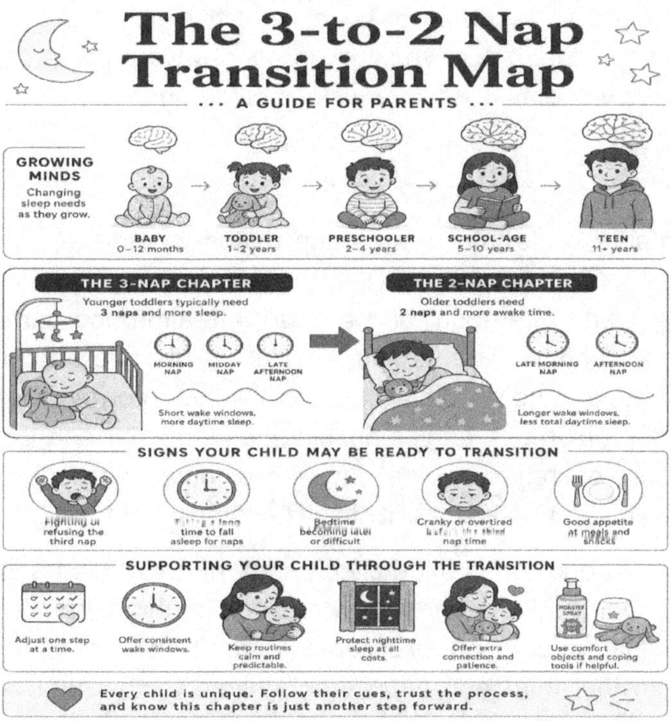

The AAP safe-sleep guidelines, designed for infants under twelve months, continue to apply for every sleep surface during this period. Daytime naps should happen on a firm, flat surface with no soft bedding, pillows, bumpers, or loose items in the sleep space. Back-to-sleep positioning remains the standard for every sleep, day and night, until your child is old enough to roll comfortably in both directions. Room-sharing without bed-sharing is recommended at least through the first 6 months and ideally through the first year. The fact that it is a nap rather than a nighttime sleep does not change the risk environment.

If your child falls asleep in a car seat, swing, or carrier, those are not safe extended sleep surfaces. Moving a sleeping baby to a firm, flat surface is the goal; briefly waking them during the transfer is a real risk, but it is the right trade-off. Many parents find that infants who nap in motion, in the car or stroller, get a lighter quality of sleep than they do in a crib or bassinet, and this partly explains why motion naps, though convenient, do not always provide the same restorative value as a stationary sleep environment.

The 3-to-2 Nap Transition

Most infants settle into a three-nap rhythm somewhere between four and six months of age. The three naps typically land in the morning, early afternoon, and a short late-afternoon catnap that bridges to bedtime. This structure works well for a few months, but somewhere around six to nine months, the late-afternoon nap starts to cause problems. It begins pushing bedtime later and later, or it disappears entirely on some days while remaining necessary on others, and the unpredictability itself becomes exhausting.

The 3-to-2 transition is often cited as the hardest nap transition of the first year, and there are a few reasons for this. First, the timing is genuinely variable. Some children are ready at six months; others hold onto that third nap happily until nine months, and both are within the normal range. Second, the signs of readiness can look almost identical to garden-variety sleep resistance or a developmental regression, so it is easy to misread what is happening. Third, the transition period itself, the two to four weeks when the third nap is sometimes needed and sometimes not, is genuinely unpredictable and requires flexibility.

The clearest signs of readiness for the 3-to-2 transition include consistent resistance to the third nap even when the wake window for it seems appropriate, early-morning waking that appears related to the late-afternoon nap pushing bedtime later, and difficulty settling at bedtime on days when all three naps happened. When you see two or more of these signs persisting for at least five days, the transition is likely real rather than a temporary disruption.

The approach that works best for most families is to allow the third nap to become progressively shorter and then move bedtime earlier to compensate. An earlier bedtime, sometimes dramatically earlier, for seven to ten days during the transition is not a permanent change. It is a short-term adjustment that prevents the overtired cortisol cycle from taking hold while the two-nap rhythm consolidates. Parents who push through the transition without adjusting bedtime often end up with a child who is overtired every evening, which disrupts nighttime sleep, shortens naps the next day, and creates a feedback loop that can persist for weeks.

Diagram 4.4: The 2-to-1 Nap Transition Map

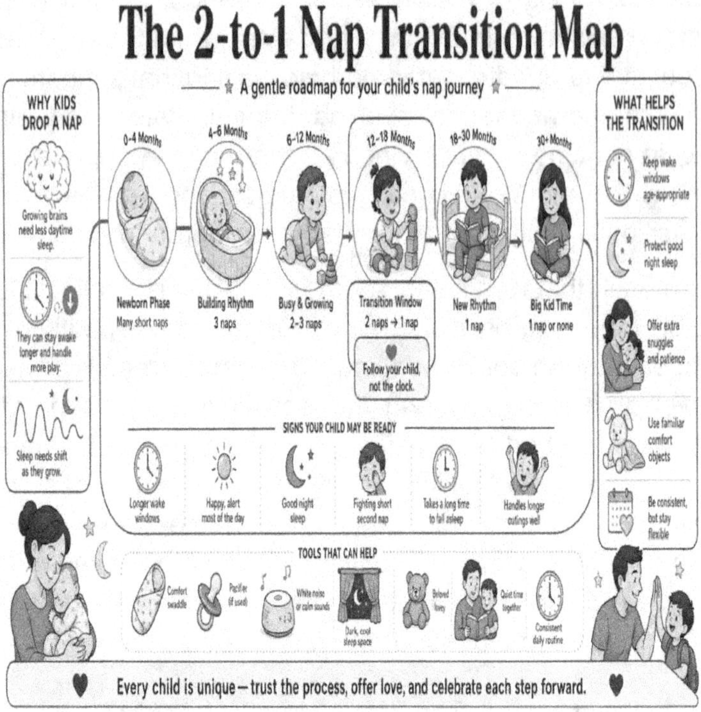

There is also a practical consideration: what happens with daycare. If your child is in a group care setting that runs on a fixed two-nap schedule, the 3-to-2 transition may occur at daycare before you decide to make it happen at home. This is not necessarily a problem. Many children handle different schedules in different environments, and daycare days and home days can differ briefly. The important thing is to recognize which environment is driving the consolidation and calibrate bedtime accordingly on each type of day.

The 2-to-1 Nap Transition

If the 3-to-2 transition catches parents off guard, the 2-to-1 transition often breaks them a little. This one tends to hit around 14 to 18 months, a stretch that is already challenging for reasons unrelated to sleep: mobility, language frustration, attachment intensity, and the early expressions of toddler autonomy are all peaking simultaneously. Adding a major nap reorganization to that mix is a lot.

The transition happens for the same biological reason as the 3-to-2: the morning nap begins to conflict with nighttime sleep. By around fourteen months, many toddlers have built enough overnight sleep consolidation that a morning nap, even a short one, is loading enough adenosine clearance to push the afternoon nap later, which pushes bedtime later, which shortens night sleep, which makes mornings harder, which makes the morning nap feel essential again. It is a circular pattern, and the only way out is through.

The signs of readiness look familiar: consistent resistance to one of the two naps, difficulty falling asleep at bedtime on days when both naps occur, and early-morning waking. Some children also show the reverse pattern, where both naps happen beautifully but the afternoon nap ends late enough to cause a two-hour bedtime battle. When you see these patterns on more days than not over two weeks, the transition is ready to happen.

The most important thing to know about the 2-to-1 transition is that the single nap is often too short at first. In the first few weeks after moving to one nap, many toddlers nap for only 45 minutes to an hour, which is not enough time for adenosine clearance for a 5-hour wake window on either side. The solution is a temporary early bedtime, usually moved up by thirty to sixty minutes, and patience. The single nap almost always lengthens within two to four weeks as the brain adjusts to the new architecture.

The single nap in this age band is typically scheduled from eleven-thirty to noon, gradually moving later as the child matures. By eighteen months, most toddlers can easily sustain a noon or one o'clock nap. Trying to maintain the nap earlier than eleven in the morning during the transition often prolongs the rocky period, because the morning wake window is not yet long enough to build sufficient adenosine for a restorative sleep.

Diagram 4.5: The 1-to-0 Nap Transition Map

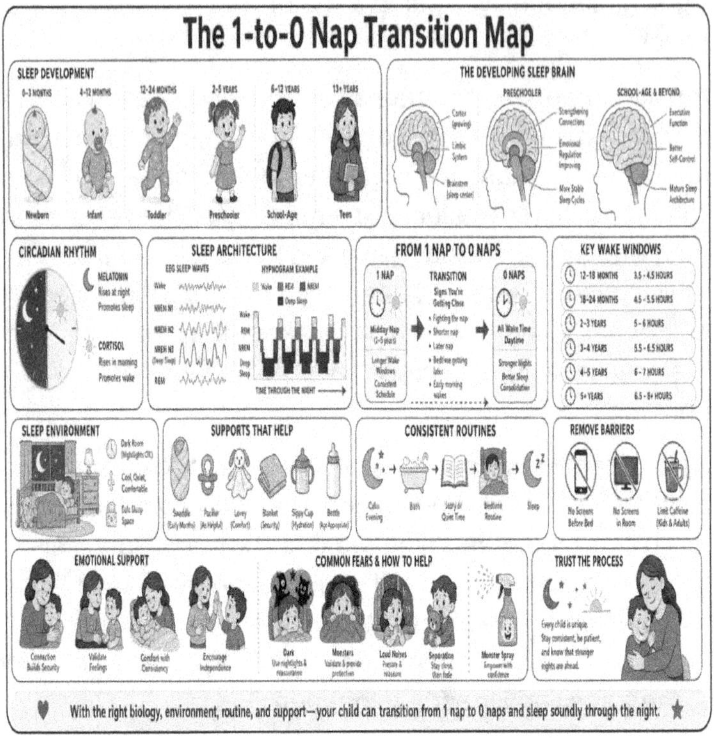

One thing that deserves acknowledgment is that this transition often coincides with daycare transitions, returns to work after parental leave, and other changes in family schedules. The logistical complexity is real. If your child's daycare still offers morning and afternoon rest periods but your child has consolidated to one nap, communicate the change clearly to caregivers and ask that your child have the option to stay awake during the morning rest and sleep through the afternoon rest. Most daycares are accommodating when the rationale is explained clearly. It is worth the conversation.

The 1-to-0 Nap Transition

This one is the most variable transition of the nap years, and also the one where parental opinion and social context play the largest role. The 1-to-0 transition, the shift from one daily nap to no nap at all, can happen as early as age two and a half in some children and as late as age five in others. The median is around age 3 to 3.5, but the range is genuinely wide, and the variation is mostly normal.

What drives the transition is the maturation of the adenosine clearance system and the lengthening of the circadian wake window. By three years, most children can sustain ten to twelve hours of waking without accumulating the kind of adenosine load that forces a daytime sleep. Their cortisol patterns have matured enough to support sustained alertness throughout the day, and their nighttime sleep has typically consolidated to the point that the overnight period provides sufficient total restorative sleep.

The signs of readiness are similar in form to those of the earlier transitions but unfold over a longer testing period. A child who consistently resists the nap for two to three weeks, falls asleep in the car frequently but not at nap time, and has stable mood and behavior in the late afternoon without a nap is likely ready to drop it. The critical test is nighttime sleep: a child who drops the nap and then crashes by five in the evening, waking up at four the next morning, is not ready yet.

The 1-to-0 nap drop transition is also more gradual than the earlier transitions. Most children benefit from a longer quiet rest period that replaces the nap, during which they are in a calm space with books or quiet toys but are not expected to fall asleep. This serves multiple purposes: it preserves a midday restorative pause in the child's day, it gives parents a predictable break, and it keeps the option open for sleep on days when the child is sick, recovering from disrupted nights, or going through a developmentally intensive period.

Diagram 4.6: Schedule vs Wake-Window vs Cue-Based Approaches

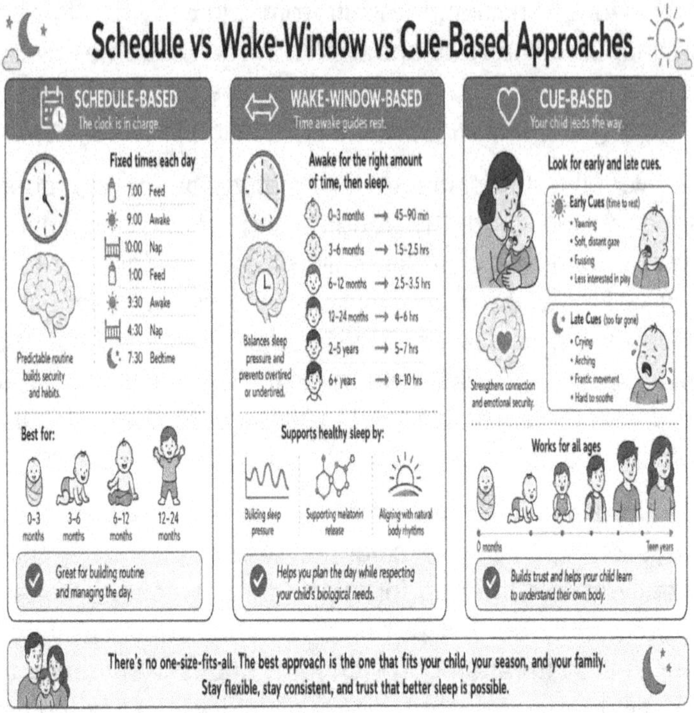

Some children will occasionally continue to fall asleep during quiet rest time, even after naps have largely dropped off, particularly on high-exertion days, sick days, and during developmental bursts. This is fine. An occasional nap at age four does not reset the nap-dropping process; it just means the child needed more sleep that day. The key thing to watch is whether the occasional nap is followed by a dramatically late bedtime or significant nighttime waking. If it is, shift quiet rest time earlier or shorten it on days when sleep occurs, to limit the impact on the night.

Schedule, Wake Window, and Cue-Based Approaches

There are three main frameworks parents use to organize the nap day, and each one reflects a different starting assumption about what matters most. Understanding all three helps you build something that actually works for your family rather than trying to follow someone else's plan.

A schedule-based approach uses fixed clock times for naps and bedtime, typically derived from a recommended framework for the child's age. The appeal is predictability: you know when naps happen, you can plan around them, and consistent timing reinforces the circadian rhythm. The limitation is that it requires some flexibility to work well. A child who slept poorly overnight may need an earlier first nap the next day. A child who had an unusually short nap may need a slightly shorter second wake window. A rigid schedule, applied without adjustment, can lead parents to put a child down who is not tired enough or to miss the window for an overtired child, and both scenarios produce nap resistance that makes parents think the schedule is not working.

A wake-window-based approach uses elapsed awake time as the cue, starting the wind-down routine when the wake window is approaching its end rather than at a specific clock time. This approach is more flexible and tends to work well for infants whose schedules shift day to day. The limitation is that it can allow schedules to drift later and later if naps run long, because the wake window starts over from whenever the nap ended. Left uncorrected, a late-running nap can push the next nap later, which pushes bedtime later, which may push morning wake time later, which starts a slow drift toward an inconvenient schedule.

Diagram 4.7: The 45-Minute Intruder Explained

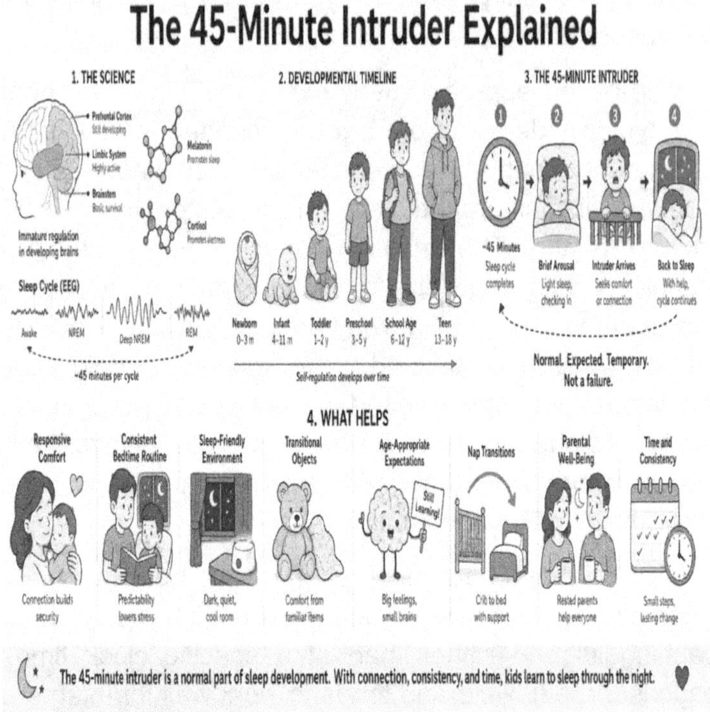

A cue-based approach watches for the child's sleep-readiness signals- the eye rubs, the yawning, the slowing down, the decrease in social engagement- and initiates the nap routine when those signs appear. This is the most instinct-driven of the three approaches and works beautifully for parents who are home and closely attuned to their child's rhythms. The limitation is that by the time sleep-readiness cues appear clearly, adenosine has already built to a fairly high level. If the nap routine takes fifteen to twenty minutes and settling takes another ten, the child may tip into overtiredness during the delay. Cue-based approaches work best when parents have learned to read the early, subtler cues rather than waiting for the obvious ones.

Most families end up using a blend: a rough schedule that provides predictability and circadian anchoring, adjusted by wake windows when the day runs differently than planned, with attention to cues as a cross-check. This blended approach has no official name, but it is what most experienced parents discover through trial and error over the first year. The good news is that you do not need to choose a single approach and stick to it. Choosing the right tool for the situation at hand is exactly the right strategy.

The phrase "perfect schedule" appears everywhere in parenting communities, and it warrants a direct response: a perfect schedule does not exist. What exists is a range of schedules that work reasonably well for a given child at a given developmental stage, and the ability to adjust when the current one stops working is more valuable than any specific schedule. Flexibility is not failure. Adjusting to your child's signals is not giving up. The parents who come through the first two years with the least sleep anxiety are the ones who learn the biology and then hold the schedule loosely.

Catnaps, False Starts, and the 45-Minute Intruder

These three phenomena are among the most universally dreaded in the nap years and among the most misunderstood. If you have ever watched a baby sleep for exactly forty-five minutes and then wake up screaming, fully confused about why they are awake, you have met the 45-minute intruder. If you have ever put a child down for a nap only to have them wake up twenty minutes later acting as though they have been fully restored when clearly they have not, you have met the catnap. And if your child has ever woken forty-five minutes into a nighttime sleep in a complete state, just as you were sitting down to dinner, you have met the false start.

Infant sleep cycles in the first few months of life are roughly 40 to 50 minutes long, shorter than adult cycles, which run about 90 minutes. Each cycle moves through phases of lighter and deeper sleep, and at the end of each cycle the brain briefly surfaces toward waking. In adults, this surface moment is almost always unconscious; adults link sleep cycles together without waking. In infants, the skill of linking sleep cycles is a developmental acquisition that typically happens over the first year.

When a baby's brain surfaces at the end of a forty-five-minute cycle and has not yet developed the ability to slide back into the next cycle independently, it wakes fully. The result is a short nap that feels like a complete nap to the infant but is not providing the deeper sleep stages of a longer multi-cycle nap. Catnaps in the first three to four months of life are essentially normal. The forty-five-minute intruder becomes more of a concern when it persists past four to five months, because by then many babies are developmentally capable of linking cycles if the conditions are right.

Diagram 4.8: Catnap and False-Start Cycle

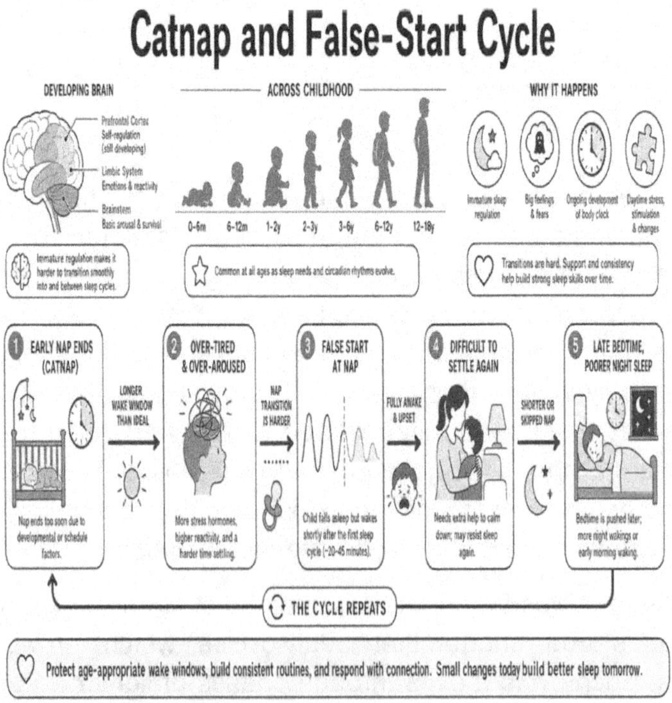

The conditions that help cycle-linking are the same ones that promote good independent sleep: falling asleep in the sleep environment where the child will remain, without a motion or feeding that is discontinued before sleep onset. A baby who falls asleep in a moving stroller and is then parked in a still stroller mid-nap has a mismatch between the sleep-onset environment and the mid-cycle environment, and the brain registers this mismatch as a signal to wake. A baby who falls asleep nursing and then wakes in a crib mid-cycle has the same mismatch. This is why sleep associations, the conditions present at the moment of sleep onset, matter so much for nap length as well as night sleep.

False starts at the beginning of the night and early-evening waking, forty-five minutes into what was supposed to be bedtime, are often driven by a different mechanism. They tend to occur when the day's total sleep was insufficient, and cortisol levels are elevated at bedtime. Cortisol disrupts the normal deepening of sleep in the first sleep cycle, causing a surface-level waking earlier than usual. The fix for false starts is almost always either moving bedtime earlier by 15 to 30 minutes, ensuring that naps were adequate that day, or both.

Catnaps in infancy (under four months) are not a problem to solve. They are a feature of immature sleep architecture and will naturally consolidate as the brain matures. Fighting catnaps in a two-month-old is a losing battle. Accepting them, planning around them, and keeping wake windows appropriate for short nap outcomes is the right response. The catnaps will get longer; the timeline is just not one you control.

Building a Flexible Daily Rhythm

The practical goal for the first two years is not a perfect schedule. It is a flexible daily rhythm that has enough predictability to anchor the circadian system and enough adaptability to survive the inevitable disruptions. Illness, travel, teething, developmental leaps, and daycare transitions will all temporarily alter the nap picture. Knowing what a good rhythm looks like for your child's current age makes it much easier to return to something functional after each disruption rather than feeling like you are starting over from scratch.

A flexible rhythm has three anchor points: a consistent morning wake time, a consistent nap window (or two, depending on age), and a consistent bedtime window. These do not need to be precise to the minute. A morning wake between six and seven, a first nap around nine to nine-thirty, and a second nap around two to two-thirty, with a bedtime between seven and eight, is a rhythm, not a rigid schedule. It gives the circadian system enough regularity to function well while absorbing the thirty- to sixty-minute variation that real family life produces.

Diagram 4.9: The Flexible Daily Rhythm

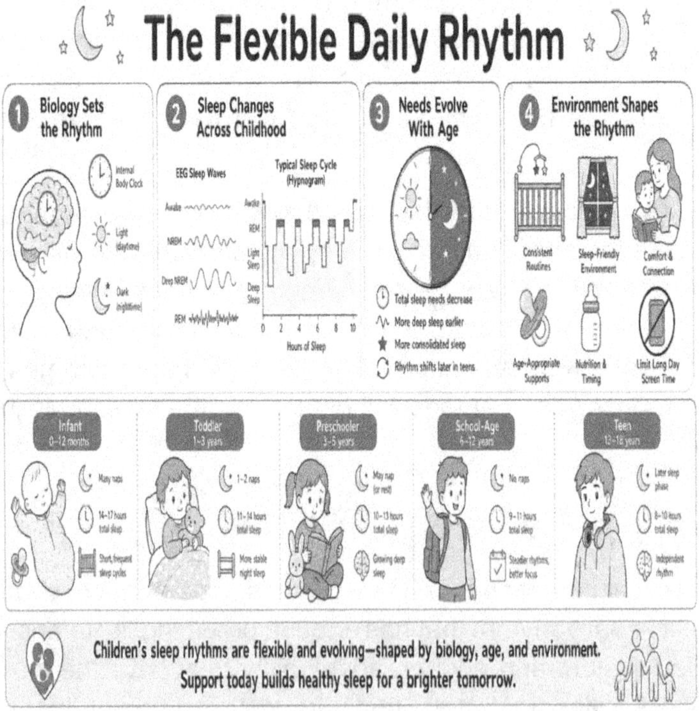

When a disruption happens, the single most useful anchor to preserve is the morning wake time. If everything else goes sideways due to travel or illness, getting your child up at the same time every morning and exposing them to natural light within the first hour is the fastest way to re-anchor their circadian rhythm and get naps back on track. Moving the morning wake time later to compensate for rough nights feels kind in the moment but usually delays the recovery of the day's schedule by another twenty-four hours.

The reality of daycare deserves its own acknowledgment here. If your child spends significant time in group care, you may have limited control over nap timing, nap length, or sleep environment during those hours. This is genuinely hard, especially during transitions. The practical approach is to communicate proactively with caregivers about where your child is in their developmental nap transition, ask for specific accommodations where possible (an earlier or later nap start, a brief wind-down before the group nap period, a wake-up if the nap runs past a certain time), and then adjust bedtime on daycare days to compensate for whatever happens with naps.

Most daycare providers appreciate parents who explain the biology rather than just making requests. Saying "she is in the 3-to-2 transition right now and the late afternoon nap is pushing her bedtime to nine o'clock" lands differently than "please don't let her nap after three." You are more likely to get a flexible, helpful response when caregivers understand the reason behind the request.

When disruptions from teething, illness, or travel resolve, give the schedule three to five days to re-establish before concluding that something fundamental has changed. Most children whose nap schedule was working well before a disruption will return to that schedule within a week of the disrupting factor resolving. Illness in particular can produce apparent nap regressions that disappear entirely once the child is well. The most common mistake is interpreting a disruption-driven change as a developmental transition and responding by dropping a nap that the child still needs.

The Overtired Loop and How to Break It

The overtired loop is one of the most important concepts in infant and toddler sleep, and it is worth explaining carefully because once you understand it, you will recognize it everywhere and be able to intervene early.

The loop works like this. A child misses a nap or has a significantly shorter nap than usual. Adenosine builds to a higher level than normal. The body responds to prolonged high-adenosine states the same way it responds to other physical stressors: by releasing cortisol and adrenaline, both of which are stimulating hormones. These hormones create a state of alert arousal that is biologically useful in a genuine stress situation but completely counterproductive for sleep. The overtired child is harder to settle for the next sleep, not easier. They take longer to fall asleep, wake more frequently during sleep, and produce shorter sleep bouts, which means adenosine clearance during that sleep is insufficient, leading to elevated cortisol levels at the next sleep opportunity and completing the loop.

The loop is much easier to prevent than to break. Prevention looks like protecting naps during transitions, adjusting bedtime earlier when nap totals are low, and resisting the temptation to keep a child up later in hopes that tiredness will translate into easier or longer sleep. Breaking the loop usually requires two to four days of consistently early bedtimes, protected and well-timed naps, and capping any single nap that is running late to preserve night sleep pressure. It is not fun, but it is finite.

Parents who recognize the overtired loop early tend to course-correct quickly and avoid the escalating cycle of ever-worse nights. Parents who interpret a child's increased difficulty falling asleep as a sign of needing less sleep tend to continue reducing sleep opportunity, which deepens the loop. Knowing the biology means you can read the difficult behavior accurately and respond to what is actually happening: a child whose nervous system is calling for more sleep, not less.

Try This Tonight

For this age range: 4 to 18 months

1. Write down your child's sleep for the past two or three days: every nap start and end time, every bedtime, and every morning wake time. Most parents discover patterns they had not consciously noticed, including wake windows that are consistently too long or too short.

2. Compare your child's actual wake windows against the approximate ranges for their age. If windows are running thirty to forty-five minutes longer than the typical range, try initiating the next sleep routine fifteen minutes earlier tomorrow and see whether settling is faster or slower.

3. Check the morning wake time. If it varies by more than forty-five minutes from day to day, experiment with a consistent anchor for five days and observe whether nap timing becomes more predictable.

4. If you are in the middle of a nap transition, move bedtime thirty to forty-five minutes earlier for the next week. A temporarily early bedtime during a transition is protective, not permanent.

5. Notice whether your child is falling asleep during naps with a feeding, motion, or other external support. If so, experiment with one nap per day that begins awake in the crib, with your normal settling support. Cycle-linking is a skill that develops with practice.

6. If daytime naps are happening in a moving stroller, car, or carrier most days, consider whether one nap per day can

happen in the crib or bassinet instead. Motion naps are lighter and shorter on average, and stationary naps provide better restorative sleep quality.

7. If your child is waking at the forty-five-minute mark of every nap, wait five minutes before going in. Some children will resettle independently. For those who do not, note the sleep association present at nap onset and consider whether that is a factor worth gradually adjusting.

Parent Takeaways

- Wake windows are the most practical tool of the first eighteen months. Learn them for your child's current age, and update them every 6 to 8 weeks as they grow.
- Naps are biologically required through at least age two and a half for most children. Dropping them early produces overtiredness, not better night sleep.
- A consistent morning wake time anchors the circadian system more powerfully than any other single scheduling choice.
- Most nap disruptions from illness, travel, teething, or development resolve within three to five days of the disrupting factor clearing. Respond with patience before restructuring.
- The overtired loop is easier to prevent than to break. Protect naps and adjust bedtime early when nap totals are low.
- Flexibility is a feature, not a failure. The best schedule is one you can actually maintain most days, with room to adjust when life intervenes.

5 The Toddler Bedtime Battle: Years 1 to 3

It is 8:17 p.m. on a Tuesday. Your two-year-old has already had three books, two glasses of water, one trip to the bathroom, one song, and one very earnest renegotiation of which stuffed animals are allowed in the crib. You are exhausted in a way that is hard to describe to anyone who has not lived it. This is the exhaustion that comes not just from being tired, but from being outsmarted, delayed, and emotionally worn down by someone who is only thirty-two inches tall.

Here is the thing: your toddler is not doing this to you. He is doing this because of how his brain is built right now. The same neurological wiring that makes him reach for your hand in a parking lot, demand to do things "by myself," and burst into tears when you cut his sandwich the wrong way is also the wiring that turns bedtime into a negotiation. It is all connected, and once you understand the biology behind it, the chaos starts to make a different kind of sense.

The toddler years, roughly ages one through three, are one of the most intense periods of brain development a human being ever goes through. Language is exploding. Motor skills are expanding. The concept of "I am a separate person with preferences and opinions" is dawning for the very first time. That dawning is thrilling and terrifying in equal measure, and your toddler has zero reliable tools for managing the big feelings it generates. What she does have, in abundance, is energy, will, and a deep attachment to the people she loves most. All of that pours straight into bedtime.

You may have had a reasonably good sleep routine in the infant months. Maybe your baby was rocking to sleep reliably, or your timing was solid. Then something shifted around twelve or fourteen months, maybe sooner, and suddenly a child who used to go down with some consistency turned into someone who needs three stories, a specific song, the nightlight at exactly the right angle, and your physical presence until she is fully unconscious. You are not imagining the regression. It is real, it is normal, and it has a very specific developmental explanation.

The goal of this chapter is not to hand you a rigid script. You know your toddler better than any book does. The goal is to give you the science behind what is happening in that tired, rapidly developing little brain, and then offer a flexible, evidence-based framework that works with his biology rather than against it. By the time you are done reading, bedtime will probably not feel effortless. But it will feel less like a battle and more like a routine you are actually in charge of.

Chapter Summary

This chapter looks at toddler sleep through a developmental lens. The biggest shift from infancy to toddlerhood is the emergence of intentional resistance: your child now has enough cognitive sophistication to anticipate bedtime, dislike the prospect, and actively work to delay it. That is not defiance in the adult sense. It is a predictable outcome of a brain in rapid development, and understanding it changes how you respond.

The chapter walks through the neuroscience of why toddlers fight sleep, how the bedtime resistance loop forms and sustains itself, what cortisol overshoot looks like in an overtired two-year-old, and why a lovey or blanket is not just comfort but actual neuroscience at work. From there, it moves into the practical: how to build a routine that genuinely calms a toddler brain, how to handle stalling with warmth and firmness, when to move from crib to bed, how to tell the difference between a nightmare and a stalling tactic, and why the tablet before dinner is costing your family more sleep than you realize.

What This Chapter Will Cover

- Why toddlers resist sleep biologically. The prefrontal cortex is the brain's braking system, and in a toddler it is still very much under construction. This section explains what that means for impulse control, emotional regulation, and the bid for autonomy that peaks in the second and third years.

- How the bedtime resistance loop forms. Toddlers learn fast. If a delay tactic works once, it gets filed away and repeated. This section walks through the loop in plain language and shows exactly how parents inadvertently train more resistance by responding inconsistently.

- Cortisol overshoot and the "second wind." Many parents push bedtime later hoping a more tired child will fall asleep faster. The science says the opposite happens. This section explains the cortisol spike that makes an overtired toddler look wired rather than sleepy, and how to work with your child's actual sleep pressure window.

- Transitional objects and the neuroscience of comfort. A well-bonded lovey or blanket is not a crutch. It is a portable piece of felt safety that a toddler can use to self-soothe in the night. This section covers the developmental psychology and the practical logistics of introducing and protecting a transitional object.

- Building a bedtime routine that actually works. Consistent, predictable, calm, and contained to thirty to forty-five minutes: these four qualities define a routine that a toddler brain can learn to trust. This section provides a flexible framework, not a rigid schedule, that you can adapt to your family.

- The "one more book" trap and how to hold loving limits. Stalling is universal. Pre-decided responses, offered with warmth and consistency, reduce negotiation to near zero. This section provides actual language you can use tonight.

- The crib-to-bed transition: timing and trade-offs. Moving a toddler out of the crib too early is one of the most common sleep mistakes in the toddler years. This section covers signs of developmental readiness, safety considerations, and transition strategies to minimize regression.
- Nightmares, night wakings, and the art of stalling. Not every nighttime appearance is the same. This section helps you tell them apart and respond in ways that are both reassuring and sleep-protective.
- Screens and toddler sleep. The mechanism by which evening screen exposure disrupts toddler sleep is specific, measurable, and worth understanding in full, because once you see it clearly, the habit change is much easier to make.

Why Toddlers Fight Sleep: The Developing Brain

To understand the toddler bedtime battle, you have to start inside the skull. At twelve months, the human brain is roughly sixty percent of its adult volume. By age three, it will reach approximately eighty percent. That is a staggering amount of growth in a very short window, and almost none of it is happening on a schedule convenient for bedtime.

The part of the brain that matters most for sleep resistance is the prefrontal cortex, the region just behind the forehead. In adults, the prefrontal cortex acts as the brain's executive manager: it weighs consequences, holds back impulses, regulates emotions, and says, "Okay, even though I want to stay up, I know sleep is good for me." In a toddler, this region is one of the last to mature. It will not be fully developed until the mid-twenties. At age two, it is barely online.

Diagram 5.1: The Immature Prefrontal Cortex

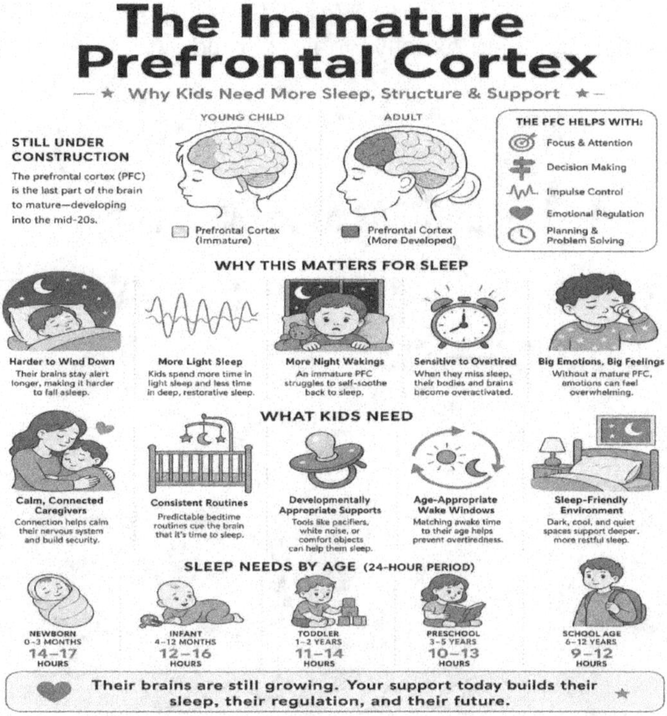

In practice, this means your toddler is not capable of the kind of rational self-regulation we sometimes expect of her at bedtime. When she says "no" to sleep, she is not being strategic the way an adult would. She genuinely cannot hold in mind the idea "I will feel better tomorrow if I sleep now" while simultaneously feeling the pull of stimulation, connection, and wakefulness. The part of her brain that would handle that trade-off is not ready yet.

At the same time, a separate developmental process is roaring forward: the bid for autonomy. Somewhere between twelve and eighteen months, most toddlers discover that they are distinct people with distinct wants. The word "no" is not just a refusal; it is a developmental milestone. It means your child has grasped the concept of self-determination. That is a genuinely important cognitive leap. It also means that being told to stop doing something fun and lie in a dark room is going to feel, to your toddler, like a profound injustice. Her drive toward autonomy is not a character flaw. It is biology doing exactly what it should.

Diagram 5.2: The Bedtime Resistance Loop

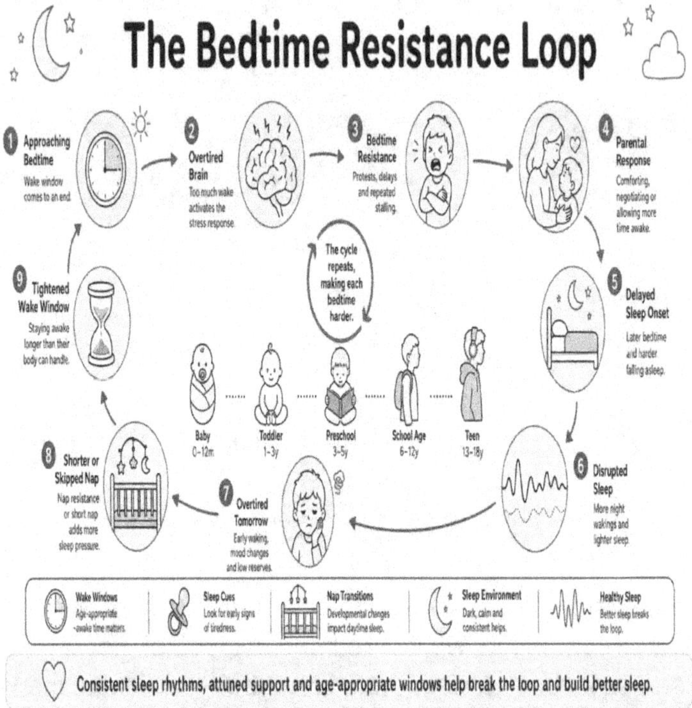

These two forces together, an immature brake system and a newly awakened drive for independence, create the perfect conditions for bedtime resistance. Your toddler wants to stay connected to you. He wants to stay in the stimulating, interesting world. He has the motor skills to get out of his crib (or, shortly, his bed). He has enough language to negotiate. And he has no reliable internal mechanism to override all of that in favor of sleep. Without a structure that does that override work for him from the outside, he will not do it on his own.

How the Resistance Loop Forms

Here is how the bedtime resistance loop works, and why it is so sticky. On night one, a toddler calls out after being put down. A parent comes in. The toddler gets comfort, connection, maybe another book. The toddler is pleased. On night two, the toddler calls out again—same result. By night seven, the toddler has learned a very reliable fact: calling out gets a parent. That fact is now encoded.

This is not manipulation in any moral sense. It is basic learning. Toddlers are extraordinarily efficient learners. They are built to notice patterns, extract rules, and apply them. The rule they extract from inconsistent bedtime responses is: persistence pays. Once that rule is in place, you will need to be very consistent for at least a week to two weeks before it begins to fade.

The hardest part of this loop is that parents reinforce it not out of negligence but out of love. Going in to comfort a crying child is a deeply instinctive response, and it is not wrong. The problem is not the going in. The problem is inconsistency: going in sometimes but not always, or going in and sometimes bringing the child to the parent's bed, or extending the routine on some nights but not others. That variable schedule of reinforcement is actually the strongest schedule for maintaining a behavior. It is exactly how slot machines work, and it is why the random one-off extra story is more powerful than a consistent extra story would be.

The fix is not to become cold or unresponsive. It is to become predictable. A warm, consistent, fully-expected response to a stalling tactic (we will get to the exact language shortly) is far more effective than an inconsistent mix of giving in and holding firm. Your toddler can relax once she knows exactly what will happen. Uncertainty keeps her brain on alert.

Cortisol Overshoot and the Second Wind

You have probably seen this: a toddler who was clearly tired at 6:30 p.m., rubbing eyes and getting clumsy, but by 7:30 p.m. has had a complete energy revival. He is bouncing on the couch. He is giggling at nothing. He looks like he could go until midnight. You think: maybe he just was not tired yet. Maybe I pushed bedtime too early. So you wait. And at 9:00 p.m., he is in a full meltdown, unable to fall asleep, taking 45 minutes to settle instead of 15.

What you witnessed is cortisol overshoot, sometimes called the second wind. It is a real physiological event.

Diagram 5.3: Cortisol Second-Wind Pattern

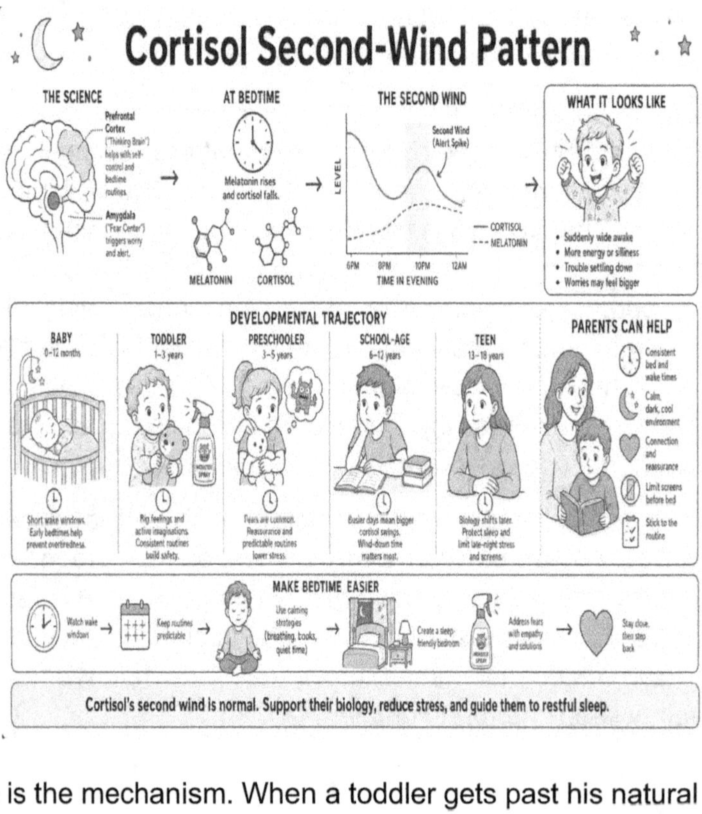

Here is the mechanism. When a toddler gets past his natural sleep pressure window (the point where his sleep drive is high and his circadian timing is aligned), his body reads the continued wakefulness as a stress signal. In response, it releases cortisol, the same stress hormone that helps adults power through a demanding afternoon at work. Cortisol is stimulating. It is supposed to fight fatigue. In a toddler, it is very effective at doing that. So your overtired toddler gets a cortisol-driven second wind that genuinely makes him look and feel more awake, even though he is running a significant sleep debt.

The cruel irony is that cortisol also makes it harder to fall asleep once you finally try. A cortisol-flooded toddler brain is activated, not ready for the easy slide into sleep. You have missed the window, and now settling will take longer, not shorter. He may finally crash, but the sleep quality will often be worse, and early waking becomes more likely because cortisol stays elevated longer than parents expect.

The practical takeaway is counterintuitive: if your toddler seems suddenly energetic and wild in the early evening, that energy spike is often the signal that he should have been heading toward sleep fifteen minutes ago, not a sign that he is not tired. Watch for the first sleepy signs (eye rubbing, decreased coordination, crankiness, quieting) and treat them as the signal to start your bedtime routine, not as a sign to wait a little longer.

Most toddlers between one and three years old have a natural sleep-onset window between 7:00 and 8:30 p.m. There is real individual variation, shaped by genetics, nap timing, and daycare schedules. The point is not to chase a specific clock time but to watch your own child's first drowsy signals and build backward from there.

Transitional Objects: The Neuroscience of Comfort

The lovey. The blanket. The worn, soft stuffed bear that has to be in the crib, or no one is going anywhere. If your toddler has one, you already know its power. If your toddler has not yet bonded to one, you might want to gently encourage it, because the science behind transitional objects is genuinely compelling.

The concept was first described by the pediatrician and psychoanalyst Donald Winnicott in the 1950s, who called them "transitional objects" because they help a child transition from total dependence on the caregiver to a growing capacity for independent comfort. Decades of developmental research have elaborated on what he described. A transitional object works because it becomes a learned signal of safety. It carries the smell of home, the texture of familiar comfort, the association of being held and loved. When a toddler clutches her blanket in a dark room after you leave, she is not substituting a piece of fabric for you. She is activating a neural pathway that her brain has learned to associate with security and calm.

Diagram 5.4: The Transitional Object Pathway

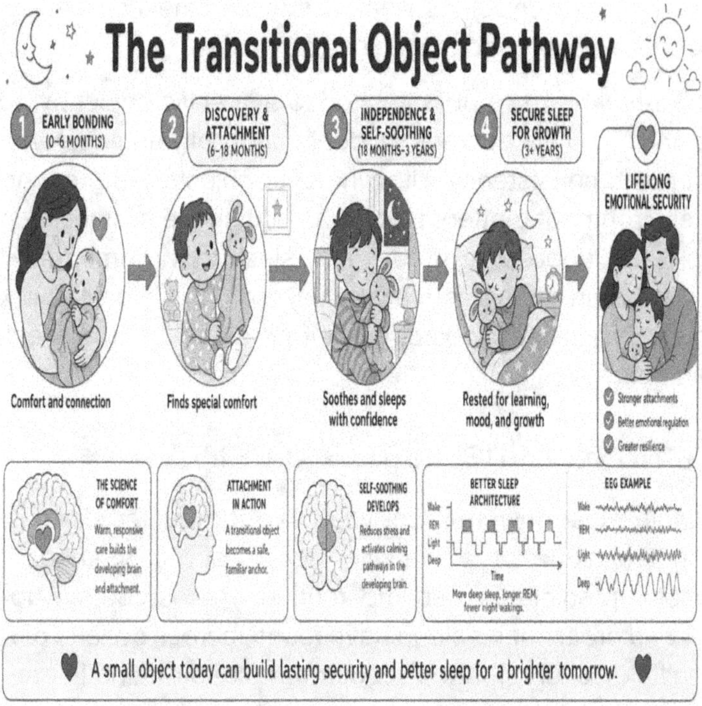

Neurologically, this works through the same basic mechanism as any conditioned comfort response. The object has been present during thousands of warm, safe interactions. Its smell and texture have been paired with the physiological state of calm and safety so many times that the brain eventually learns to produce that calm state in response to the object alone, even without the caregiver present. This is not a workaround or a trick. It is genuine self-soothing capability that you have helped your child build.

For toddlers under two, introducing or reinforcing a transitional object is fairly simple. Keep the object close during feeding or nursing, bring it into warm moments throughout the day, and consistently place it in the crib at sleep time. If your child has already chosen an object, protect it fiercely. Have a backup if the original is ever lost. Wash the object occasionally but not compulsively; part of its comfort power comes from its familiar smell, and a too-clean lovey is a lovey stripped of its signals.

For children who have not attached to a specific object by age two or two and a half, it becomes harder to artificially engineer the bond. At that point, consistency with whatever objects are present in the sleep environment matters more than finding the perfect lovey. The principle still applies: a predictable sleep environment, with the same objects in the same place every night, functions as a kind of distributed transitional object even if no single item has achieved totemic status.

Bedtime Routines That Actually Work

The single most powerful tool you have for toddler sleep is not a specific technique or a specific sleep training method. It is a consistent, predictable bedtime routine. Researchers who study toddler and preschooler sleep have found, across dozens of studies and multiple cultures, that a regular pre-sleep routine is one of the strongest predictors of faster sleep onset, fewer nighttime wakings, and better mood the following day. Not a perfect routine. A predictable one.

Here is why predictability matters so much at this age. A toddler's brain is still learning to trust the world. The prefrontal cortex cannot yet talk a toddler down from anxiety in the abstract way an adult can reason through their worries. What does work is pattern recognition. If the same sequence of events happens every single night, the toddler brain learns to use that sequence as a signal: this is what sleep coming feels like. The sequence itself becomes calming, not because each element is inherently sedating, but because the brain learns to associate the sequence with the physiological state of winding down.

Diagram 5.5: A Toddler Bedtime Routine That Works

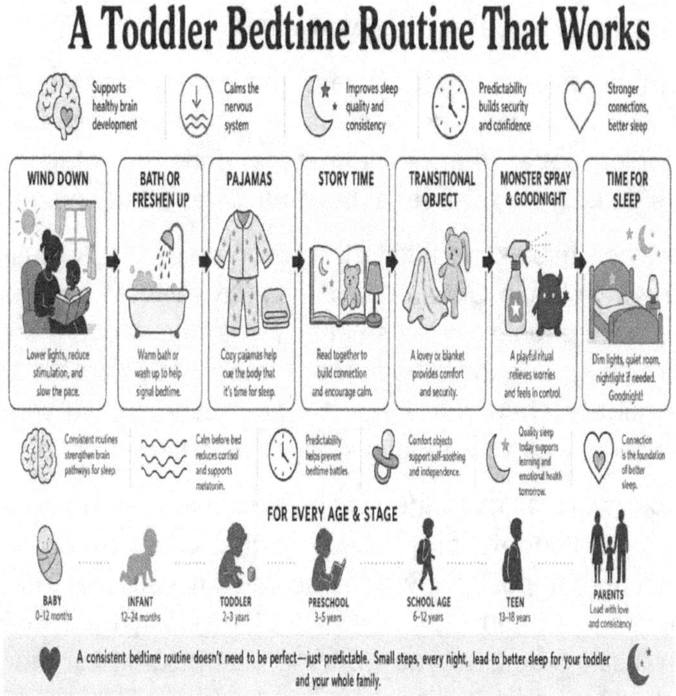

A good toddler bedtime routine has four qualities. It is consistent, meaning it happens in the same order every night. It is predictable, meaning your toddler can anticipate what comes next (and will, eventually, start completing the steps before you prompt them). It is calming, meaning each step reduces rather than increases physiological arousal. And it is contained, meaning it fits inside a thirty-to-forty-five-minute window and has a clear, agreed-upon end.

A typical routine for a two-year-old might look something like this. It begins with a bath or a warm washcloth face-wash, which has a genuine physiological effect: the drop in body temperature after warmth accelerates the shift toward sleep. Then pajamas, chosen (within limits) by the toddler. Then teeth brushing, which is non-negotiable and has a fixed two-minute span. Then one to two books in a specific reading chair or on the bed. Then the lovey goes in, the light goes off or to nightlight level, and a short song or a simple goodnight ritual happens. Then you leave.

The number of steps matters less than the consistency of the sequence. Four steps done in the same order every night will serve your toddler better than seven steps done in rotating order. Toddlers do not need variety at bedtime. They need certainty.

One common mistake is letting the routine expand. A routine that was thirty minutes in October becomes fifty minutes by December because each expansion felt small in the moment: one more book, one extra song, one more back rub. Watch for this creep, and trim back decisively when you notice it. The routine should serve your toddler's sleep, not your toddler's negotiating talent.

A note on choices within the routine. Toddlers are in the midst of discovering autonomy, and a bedtime that offers no choices can feel like a pure imposition, which escalates resistance. The trick is to offer choices about non-essential elements while keeping the structure itself fixed. "Do you want the blue pajamas or the striped ones?" is a real choice. "Do you want to do the routine tonight or not?" is not a choice you offer. The choices feel meaningful to your toddler, satisfy the autonomy drive, and cost you nothing in terms of the structure that actually matters.

The One More Book Trap

Every parent of a toddler has been in this moment. You have finished the agreed-upon books. You have said the goodnight words. You are at the door. And a small, hopeful voice says, "One more book?" The eyes are large. The voice is soft. You love this child more than you knew it was possible to love anything. And the book only takes seven minutes.

You give the extra book. The next night, the same request arrives a little earlier in the routine because your toddler now knows it works. Within two weeks, "one more book" is an established expectation, and when you finally say no on a night when you genuinely cannot, the resulting meltdown is ten times more intense than it would have been if you had held the line on night one.

Diagram 5.6: The One-More-Book Trap

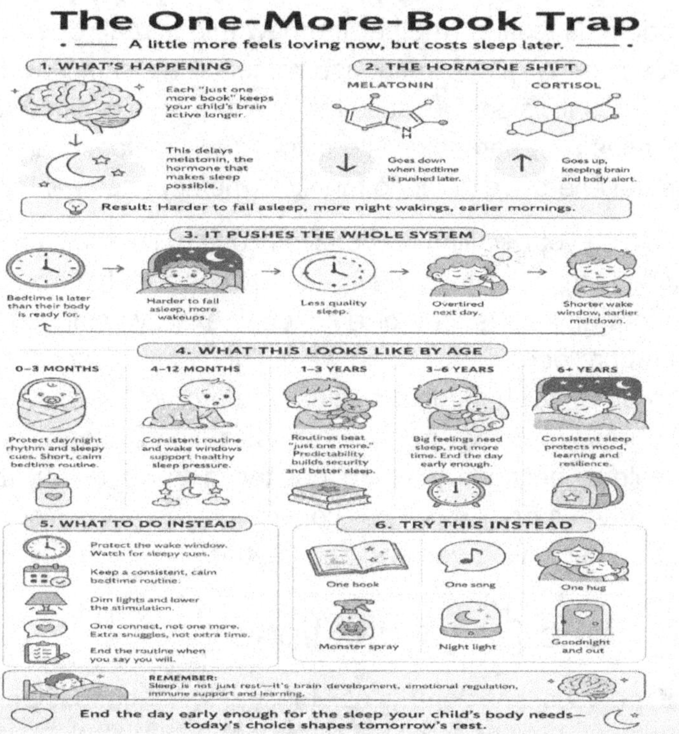

This is not a discipline problem. It is a learning problem, and the learner in question is very good at her job. The solution is not to become inflexible or unkind. It is to build the boundary into the routine from the beginning, so clearly that the negotiation never really gets off the ground.

Before the routine starts, say clearly and calmly: "Tonight we are going to read two books." Hold up two fingers if it helps. When the two books are finished, close the last book and say, "That was our two books. Now it is time for the goodnight song." Do not pause at the door for negotiation. Do not offer a lengthy explanation of why no more books tonight. The response to "one more book" is warm, short, and utterly consistent: "We read our books. It is time for sleep. I love you. Goodnight."

The first few nights of holding this line will likely involve protest. That protest is not evidence that your limit is wrong. It is evidence that your toddler is testing whether the new rule is real. She will test it more than once. If the response remains the same every time, testing gradually stops because the rule has been confirmed stable. Stable rules are, paradoxically, calming to toddlers. They reduce the mental energy required to monitor whether rules will hold.

Pre-deciding your stalling responses is one of the most underrated parenting tools. When you have already decided, before 7:00 p.m. and before anyone is tired or emotional, exactly what you will say to "one more book," to "I need water," to "I am scared," and to "I need to go to the bathroom again," you are not making decisions under duress at 8:15 p.m. You are executing a plan. Your voice stays calmer because you are not trying to figure out the right answer at the same time. Your consistency goes up. Your toddler gets clearer signals. Everyone does better.

The Crib-to-Bed Transition

One of the questions parents ask most often in the toddler years is when to move a child from the crib to a "big kid" bed. The intuition is often that the transition should happen early, sometime in the second year, because it feels like a developmental milestone to offer. The research and the clinical experience both suggest a different answer: wait longer than you think.

Most children do best staying in the crib until between two and a half and three years old, as long as they are not climbing out of it unsafely. The crib does something important that a toddler bed cannot do: it provides a physical boundary that mirrors the psychological boundary you are trying to establish at sleep time. A toddler who cannot get out of her sleeping space is much more likely to stay in it. A toddler who can get out will. Repeatedly. For months.

Diagram 5.7: Crib to Bed Transition Decision

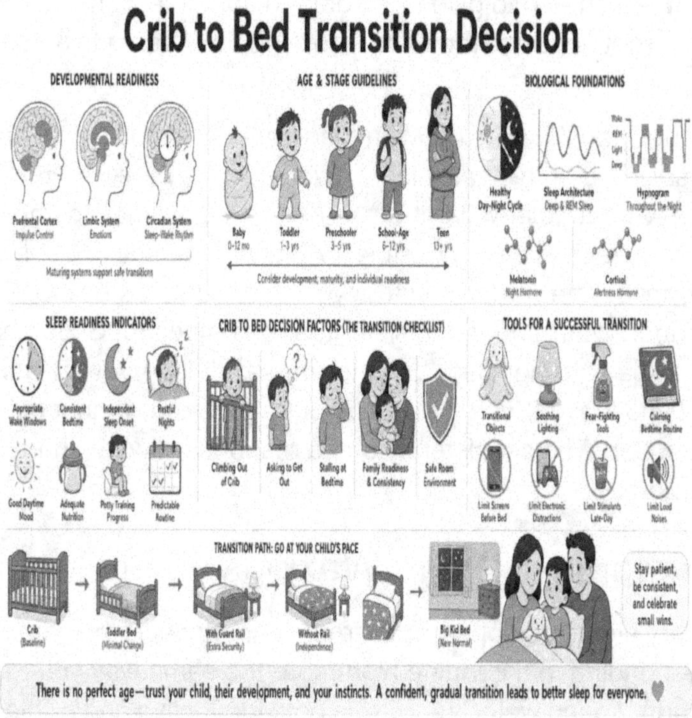

The genuine signal to move is not age, sibling pressure, or parental readiness. The genuine signal is one of the following: the child can climb out of the crib and is at risk of injury; the child is approaching 3 years old and is demonstrating the self-regulation to stay in bed with minimal boundary support; or a second baby is arriving, and the crib is needed. Even in the last case, it is worth borrowing or buying a second crib rather than rushing the transition before the older child is ready.

When you do make the move, do it gradually and with ceremony. Let your toddler help choose the new bedding, if possible. Keep the new sleep environment as similar as possible to the crib environment: same room, same lighting, same lovey, same routine. Consider using a toddler rail on one side of the bed so the transition to an open mattress is not abrupt. Some families find it helpful to use a visual cue (a simple nightlight on a timer, or a clock that shows a sun at wake time) to give the toddler a concrete signal for when it is okay to get up, since the word "morning" is abstract to a two-year-old.

After the transition, expect a regression period. It is almost universal, and it typically lasts two to four weeks. During this window, your toddler will get out of bed multiple times, come to find you, and test every possible variation of the post-bedtime appearance. The response strategy is the same as it is for stalling at the door: warm, brief, consistent, and immediate. Scoop her up, walk her back to her bed, tuck her in, say "I love you, it is time to sleep," and leave. Every time. Without anger and without extended conversation. The consistency of the response is what shortens the regression.

Nightmares vs Night Wakings vs Stalling

Not every nighttime appearance from your toddler is the same thing, and responding to a genuine nightmare the same way you respond to a fourth "I need water" request is both ineffective and a little unfair to your child. Learning to tell these apart takes some practice, but the patterns are recognizable.

A genuine nightmare typically happens in the second half of the night, when REM sleep is heaviest. Your toddler will wake abruptly and be genuinely distressed: crying, calling for you, sometimes unable to settle quickly. She may or may not remember what scared her. She may want to describe something frightening, though at two the description is likely to be fragmentary ("there was a big one"). The appropriate response is warm and present: go in, offer physical comfort, speak calmly ("You had a scary dream. It was just a dream. You are safe. I am here."). Stay until she is calm but not until she is asleep. Gradually shorten the stay over subsequent nights as she reestablishes comfort in her space.

Diagram 5.8: Toddler Stalling Patterns

Night terrors are different from nightmares and worth naming here. A night terror typically happens in the first third of the night, during deep slow-wave sleep. Your child may be screaming, sitting up, and appearing distressed, but she is not awake. Her eyes may be open and unseeing. She will not respond to your comfort in the way she would if she were awake. She will not remember anything in the morning. The counterintuitive guidance for night terrors is to do very little: keep her safe, do not try to wake her, speak softly from nearby, and wait for it to pass. Trying to hold or comfort a child in a night terror can sometimes intensify it. Night terrors tend to run in families, are more common in overtired children, and usually resolve by the preschool years.

Plain night wakings, as opposed to nightmares or terrors, are wakings that happen because of a learned association. If your toddler learned to fall asleep with you in the room, she will likely wake at night looking for that presence when she surfaces between sleep cycles (which all humans do). She is not scared. She is not in distress. She is recalling the conditions under which she originally fell asleep and requesting them. The response here is different: brief, warm, matter-of-fact, and consistent, gradually fading your involvement over two to three weeks rather than rushing to extinction.

And then there is stalling. The hallmark of stalling is the specific, targeted, escalating request: water, bathroom, one more kiss, a question about something clearly non-urgent. The requests often come in sequence, each one reasonable-sounding enough to generate a response. A toddler who has stalled effectively for three nights running will generate more sophisticated and sympathetic-sounding requests over time, because the effective ones get reinforced and the ineffective ones drop away. This is not cunning. It is learning. And the response, again, is warm and consistent: "I love you. It is time to sleep." Followed by leaving.

Why Screens After Dinner Wreck Toddler Sleep

It is worth spending some time on this one, not to make you feel guilty about screen time (most families use it, it is not inherently wrong, and you are doing your best), but because the mechanism here is specific and, once you understand it, it genuinely changes your calculus.

Screens emit light across the visible spectrum, but the wavelength that matters most for sleep is short-wavelength blue light, roughly 460 to 480 nanometers. That particular band of light is what the circadian system is most sensitive to. Special photoreceptors in the retina called intrinsically photosensitive retinal ganglion cells, or ipRGCs, are maximally responsive to exactly this range. When those cells detect blue-range light, they signal the suprachiasmatic nucleus (the brain's master clock), which responds by suppressing melatonin production.

Diagram 5.9: Screen Light and Toddler Sleep

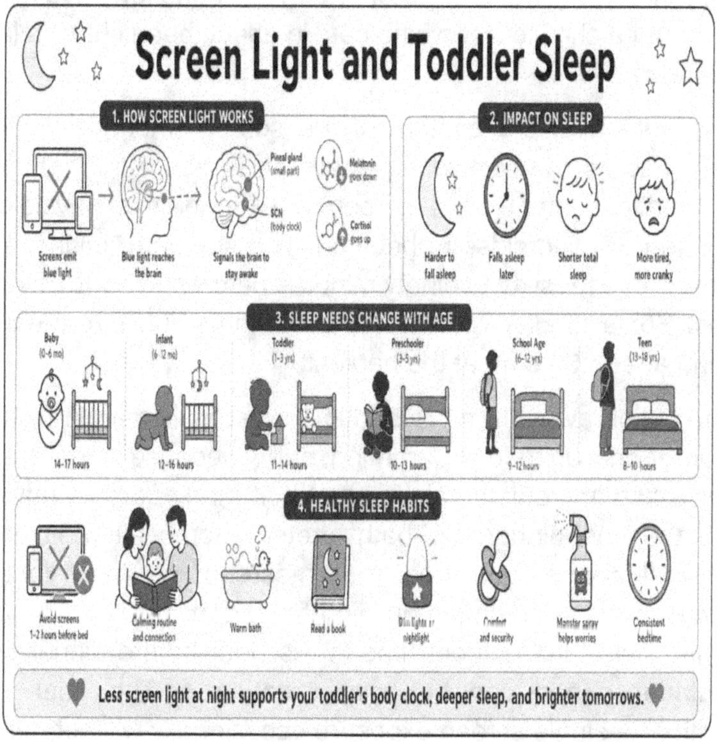

In adults, two hours of evening screen exposure has been shown in controlled studies to delay melatonin onset by roughly 90 minutes and reduce melatonin concentration by about 20%. The adult studies get cited frequently. What gets less attention is that the toddler retina has larger pupils relative to adults and may transmit more light per unit of exposure; some researchers believe toddler sensitivity to evening light suppression is even greater than adult sensitivity, though the exact magnitude is still being studied.

Here is what that means in practice. If your two-year-old watches a forty-five-minute show on a tablet or television between 6:30 and 7:15 p.m., the blue light from that screen is actively suppressing the melatonin that his body needs to start releasing for him to feel sleepy by 7:45 or 8:00. He is then placed into his bedtime routine in a neurochemically wakeful state. Then you spend forty-five minutes trying to get a child to sleep who has, in effect, been chemically told to stay awake.

This is not about the content of the screen (though stimulating content has its own activating effect). It is about the light. Even calm, gentle content on a screen will trigger this physiological response if viewed close to bedtime. The standard guidance is to avoid screens for sixty to ninety minutes before the bedtime routine begins. For a toddler with a 7:30 p.m. lights-out, that means no screens after 6:00 p.m. at the earliest.

The transition away from pre-bedtime screens is genuinely one of the harder ones in toddler parenting, partly because screens are an effective and immediate calming tool that many parents rely on to bridge the post-dinner, pre-bath stretch. Some alternatives that work at this age: audiobooks (non-screen, no blue light, still calming), puzzles or simple building toys, drawing or coloring at a low-light table, and outdoor time in the hour before dinner so the post-dinner period can be calmer naturally. The goal is not perfection. A single screen exposure occasionally will not ruin your toddler's sleep. A nightly pattern of screen time in the sixty-to-ninety minutes before bed probably will delay sleep onset by enough to shorten total sleep time in a meaningful way over weeks and months.

Holding Loving Limits

One phrase that runs through this chapter is worth naming explicitly: loving limits. The toddler years are a time when children deeply need both the warmth of connection and the security of consistent limits. The research on this is not subtle. Toddlers in environments with warm, consistent limit-setting show better sleep, better emotional regulation, and lower rates of anxiety than toddlers in environments where limits are either absent or inconsistently applied.

This does not mean rigidity. It means that when you say a routine will have two books, it has two books. When you say goodnight at the door, you say goodnight. When your toddler calls out for the fourth time, your response is the same as it was for the first time: warm, brief, and clear. You can be gentle and still be consistent. You can be loving and still leave the room. In fact, the consistency of your limit is one of the ways you communicate love, because it tells your child: the world is stable, I am safe, I can relax.

Parents in the toddler years often oscillate between firmness and giving in, not because they do not know what they want to do but because it is so hard to hold a limit when a child is crying, and you are exhausted at 8:30 p.m. That oscillation is understandable. But it is also the oscillation that makes the problem worse over time. The investment of one to two weeks of genuine, consistent follow-through pays back months of easier bedtimes. That arithmetic is worth sitting with.

You are not being unkind by leaving the room after the agreed-upon routine. You are not abandoning your child by not returning for the fourth glass of water. You are doing something that is genuinely hard and genuinely important: you are teaching your toddler that the nighttime is safe, that the routine is real, and that sleep is a place she can go without needing to fight her way there.

Try This Tonight

For this age range: 1 to 3 years

1. Write out your current bedtime routine in exact sequence, with the clock time each step starts. If the total time from the first step to lights-out exceeds 45 minutes, identify the two steps that can be shortened or removed, and plan to trim them tomorrow night.

2. Pick the two stalling tactics your toddler uses most reliably. Write down, in your own words, the exact short response you will give to each one tonight. Practice saying it out loud before 7:00 p.m. so it does not have to be improvised at 8:15.

3. Check the timing of any screens tonight. If your toddler watches anything in the ninety minutes before the routine starts, plan a one-week experiment of moving that viewing earlier and note whether settling takes less time.

4. If your toddler does not have a consistent lovey or comfort object, place a soft, small stuffed animal in the crib or bed tonight and keep it there every night. This is a long-game strategy, but it starts on night one.

5. Notice the first evening sleepy signal, the eye rub, the clumsiness, the quieting, and note the clock time. That is your actual bedtime window. If your routine currently starts thirty minutes after that signal, experiment with starting it at the signal instead.

6. If you have been going back in repeatedly after lights-out, plan to use one return tonight: a single brief, warm check-in if calling escalates, then no more. The goal is not to ignore your child. The goal is to reduce the variability in your response, which worsens the resistance.

7. If you are near or past the two-and-a-half-year mark and the crib is still working well, do not rush the transition. If you are actively considering the move, spend this week watching for the

three readiness signals mentioned in this chapter before deciding.

Parent Takeaways

• Toddler bedtime resistance is not defiance. It is a predictable outcome of an immature prefrontal cortex and a newly activated drive for autonomy. Understanding the biology makes it easier to respond calmly.

• Consistency, not perfection, is what matters in a bedtime routine. A four-step routine done in the same order every night is more valuable than a seven-step routine done each time differently.

• Pre-decide your stalling responses before bedtime begins so you are not negotiating in real time when you are tired. Warm, brief, and consistent is the formula.

• Cortisol overshoot means that an overtired toddler looks wired, not sleepy. Catching the first drowsy signal and starting the routine then is more effective than waiting for a more obvious tired cue.

• Screens in the sixty-to-ninety minutes before bed suppress melatonin and delay sleep onset. Even a one-week screen-free pre-bedtime experiment will usually show a measurable difference in settling time.

• The crib-to-bed transition is usually better later than you think. Most children do well in a crib until close to three, unless they are climbing out of it unsafely. Waiting until genuine readiness reduces the regression window significantly.

6 Preschool Sleep: Ages Three to Five

It is 5:47 on a Tuesday evening, and you are standing in your kitchen trying to get dinner on the table. Your four-year-old is on the floor behind you. Not hurt. Not sick. Just completely undone. The crime, as best you can piece together, is that you poured her apple juice into the blue cup instead of the green one. She is crying the way people cry when something enormous has happened. And somewhere in the back of your brain, through the fog of your own long day, you know exactly what is going on: she skipped her nap at preschool again.

You remember the version of her from two hours ago, cheerful and chatty in the pickup line, telling you about the caterpillar she found on the playground. That child seems very far away right now. The child on the floor behind you is operating on fumes, and no amount of reasoning, redirecting, or cup-swapping will fix that in the next four minutes. You know this. But it does not make it easier to stand there and breathe through it.

Here is what is also true: this phase, the messy, unpredictable, she-naps-sometimes-and-sometimes-does-not phase, is one of the hardest stretches in the whole preschool sleep story. It is not a failure of your routine. It is not your child being difficult to punish you. It is a genuine neurological transition, one that unfolds over months, sometimes years, and that looks different in every child. Some kids drop the nap cleanly, as a switch flipped. Others go back and forth for what feels like forever, napping at preschool on some days and refusing entirely on others. Both are normal. Both are survivable.

The preschool years, roughly ages three through five, are a time of enormous developmental change. Your child's brain is wiring up at a furious pace, building language, imagination, emotional regulation, and social skill, all at once. Sleep is not separate from that growth. Sleep is actually where a huge part of that wiring happens. Deep slow-wave sleep in the early part of the night consolidates the day's learning. REM sleep, concentrated in the second half of the night, processes emotional experience. Both are essential. And both are shifting, maturing, and reorganizing in ways that directly shape what bedtime looks like in your house.

What shifts alongside the neuroscience is imagination. The same astonishing cognitive leap that lets your four-year-old narrate a twenty-minute story about a dragon who owns a bakery also lets her lie in the dark at 8 p.m. and become genuinely convinced that something is under the bed. The monster is not metaphorical to her. The fear is real. Knowing how to respond to that, how to take it seriously without making it bigger, is one of the core parenting skills of this age window.

This is the chapter for you if you are in the thick of any of it—the nap drop. The 9 p.m. curtain calls. The nightmare at 2 a.m. The night terror at 11 where she screamed and seemed awake but was not. The bedtime fear that escalates into an hour-long production. You are not alone in any of it. And there is a clear path through.

Chapter Summary

This chapter covers the preschool window, ages three to five, a stretch defined by real cognitive sophistication, the gradual loss of the daytime nap, and the first appearance, in many children, of true sleep-onset anxiety, nighttime worries, nightmares, and night terrors. Preschoolers can articulate fears in ways toddlers cannot. They can also negotiate, delay, and escalate at bedtime in ways toddlers cannot. Both of those facts matter for how you structure sleep.

The neuroscience is also shifting in significant ways. By age five, sleep architecture and circadian timing are quite close to adult patterns, even though total sleep need is still higher than an adult's. Slow-wave sleep deepens and consolidates. REM cycles lengthen.

And the pressure that drives parasomnias, those confusing, sometimes frightening events like night terrors and sleepwalking, is at its peak during these years. Understanding what is happening in your child's brain during these events transforms your response from panic to calm competence.

What This Chapter Will Cover

- The nap-drop window, ages three to five. Most children drop the nap somewhere in this range, and almost none need it by the time they start kindergarten. Quiet time is the healthy replacement. An earlier bedtime fills the gap left by the nap. Knowing how to read your child's cues makes the transition much smoother.

- Preschool sleep architecture. The cycles are getting longer, slow-wave sleep is getting deeper, and the distribution of sleep stages across the night is maturing. Understanding the shape of your child's night helps you make sense of everything from early waking to late-night terrors.

- Nightmares versus night terrors. These are two completely different neurological events, and they require opposite parental responses. Nightmares occur during REM sleep, happen late at night, and your child is fully awake and remembers them. Night terrors come from deep non-REM sleep, happen in the first third of the night, and your child is not fully conscious and will not remember them in the morning. Responding correctly to each one matters.

- Sleepwalking and confusional arousals. Night terrors are related to these events and belong to the same family of non-REM parasomnias. They are common in preschoolers, rarely signal any medical problem, and are mostly managed through safety and patience.

- The role of imagination at bedtime. Preschoolers have rich, active inner lives and the cognitive ability to generate genuinely frightening mental images. Monster spray, nightlights, comfort objects, and brief calm acknowledgment are all legitimate tools. Extended bedtime worry-processing is not.

• Starting preschool or pre-K. New schedules, new stimulation, new social stress, and sometimes a new nap environment all converge when a child enters group care. Sleep often worsens before it improves during this transition, and there are practical strategies for managing the disruption.

• Two-parent consistency and the bedtime negotiation cycle. Preschoolers are developmentally primed to find and exploit inconsistency between caregivers. Alignment between adults is not just helpful; it is the single biggest structural factor in whether bedtime runs smoothly or becomes a nightly standoff.

Sleep Architecture at Ages Three to Five

If you have been following this book from the beginning, you have watched your child's sleep architecture change dramatically since birth. The forty-five-minute newborn cycle, the fragmented toddler night, the unpredictable nap schedule: all of it has been moving, gradually but steadily, toward something that looks more like adult sleep. By the time a child reaches the preschool years, that maturation is well along, though not complete.

Adult sleep cycles run roughly ninety minutes. Preschoolers are still running cycles that are slightly shorter, closer to sixty to seventy-five minutes on average. Still, they are meaningfully longer than the forty-five-minute cycles of infancy. What this means practically is that your three-year-old has fewer total transitions between cycles during the night than your one-year-old did. Each transition is still a moment when brief arousal is possible, but the overall architecture is more consolidated and more stable.

Diagram 6.1: Preschooler Sleep Architecture Across the Night

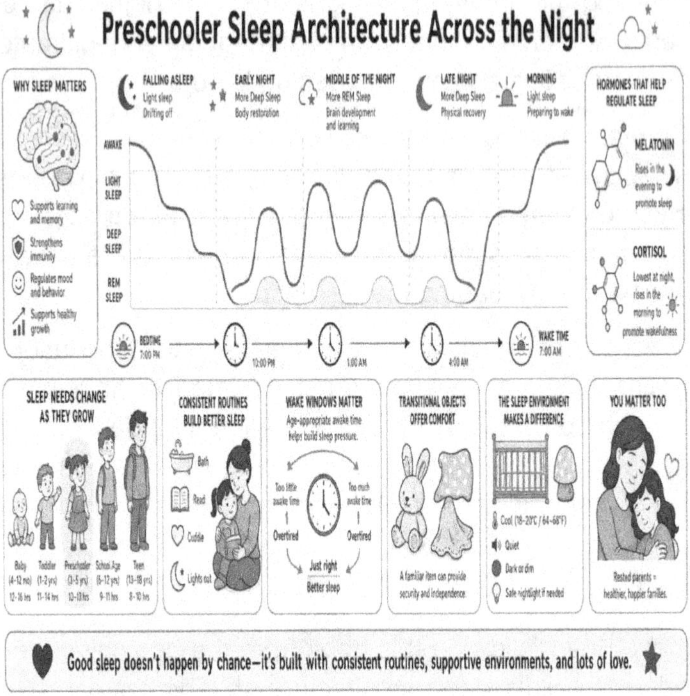

Slow-wave sleep, sometimes called deep sleep or Stage N3, is at its absolute lifetime peak during the preschool and early school-age years. Children this age produce more slow-wave sleep, both in absolute time and as a proportion of total sleep, than they will at any other point in their lives, including infancy. This is not incidental. Slow-wave sleep is where growth hormone is primarily secreted. It is where the brain consolidates the procedural and declarative memories laid down during the day. It is where physical restoration happens most intensely.

The flip side of all that deep slow-wave sleep is that waking a child from it is genuinely hard. Slow-wave sleep is the sleep stage from which it is most difficult to wake someone, and when arousals do happen, partial arousals are common. A partial arousal from deep slow-wave sleep is what produces night terrors, sleepwalking, and confusional arousals. The same biology that makes your preschooler such a sound sleeper in the first few hours of the night is the biology that drives those dramatic, disorienting events. There is no contradiction there. It is all the same system.

REM sleep, the stage associated with vivid dreaming, increases in the second half of the night as slow-wave sleep pressure dissipates. By the preschool years, children are having rich, narrative dreams. The narrative capacity of their waking imagination maps directly onto the narrative complexity of their dreams. This is why nightmares in this age group are genuinely frightening and can feel very real for minutes or even longer after waking.

Total sleep need for a three-year-old is typically eleven to fourteen hours over a twenty-four-hour period. By age five, that range shifts slightly downward, to ten to thirteen hours. These numbers include any daytime napping. A five-year-old who naps for an hour in the afternoon may naturally wake earlier in the morning or resist bedtime longer at night. The math is honest. When you add sleep in one place, it often subtracts from another.

The Nap-Drop Window

There is a reason parents talk about nap drop with something close to grief. The nap is not just sleep. It is a predictable rest point in your day, a window when the house is quiet, and you can eat lunch standing over the sink in peace, fold laundry, or sit for ten minutes. When that window starts to shrink and eventually disappear, it genuinely changes your day.

Most children drop the nap somewhere between ages three and five. Research suggests the average is around age three and a half, but that average conceals enormous variation. Some children reliably nap until their fifth birthday. Others stop napping consistently before their third. Neither extreme is a sign that something is wrong. What matters is reading the signals your individual child is giving you, not matching a timetable.

Diagram 6.2: The Nap-Drop Window, Ages Two to Five

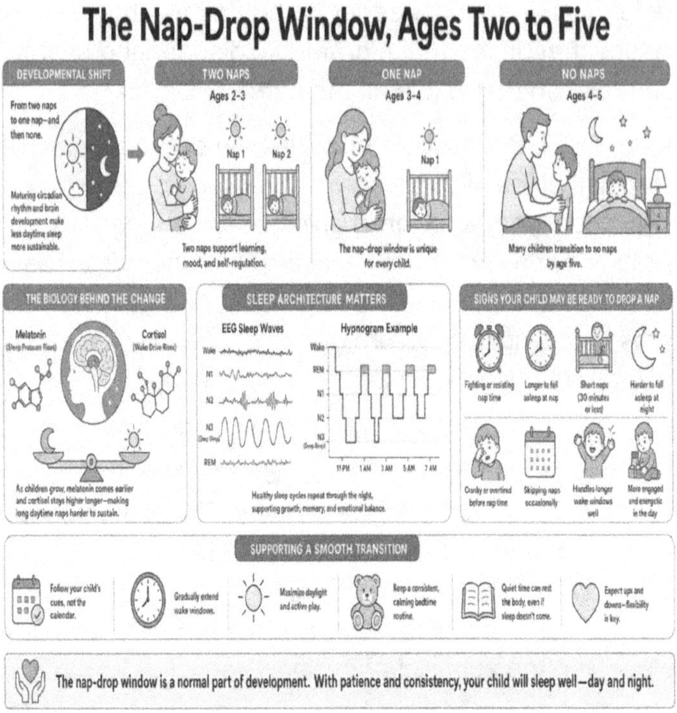

The clearest signal that a child is ready to drop the nap is that nap refusal coincides with no observable increase in nighttime sleep. If your child stops napping and replaces that hour with more consolidated, uninterrupted night sleep, the nap was ready to go. If your child stops napping and becomes progressively more overtired, harder to settle at night, wakes more frequently, or melts down by late afternoon, the nap probably still serves a purpose, and you are in the resistance phase rather than the true drop phase.

A second signal is what happens to bedtime. Children who are genuinely done with napping can handle an earlier bedtime without difficulty. Moving bedtime from 7:30 to 6:30, or even 6:00, on nap-free days is not unusual and often results in a child who falls asleep faster, sleeps better, and wakes at a normal time in the morning. Parents frequently resist this, convinced that an early bedtime means an early waking. For most preschoolers, the opposite is true.

An overtired child is a child with elevated cortisol and adrenaline who will have more trouble falling asleep and more fragmented sleep through the night.

A third signal, less talked about but worth knowing, is nap-drop ambivalence. On some days your child naps beautifully, sleeps for ninety minutes, and wakes cheerful. On other days the same child rolls around in bed for an hour, never sleeps, and is fine. This phase, where napping is optional rather than essential, can last for months. You are not doing anything wrong. You are in the middle of a genuine biological transition, and your child's body has not fully resolved it yet.

The practical strategy for nap-drop ambivalence is to offer a rest time rather than a mandatory nap. Put your child in a calm, dimly lit space, provide a few quiet activities (books, soft toys, audio stories), and step back. Some days sleep happens. Most days it does not. Either outcome is acceptable. The rest itself has value even without sleep: cortisol levels drop, arousal decreases, and the afternoon becomes more manageable for everyone.

Quiet Time as a Bridge

Quiet time is not a consolation prize for a missed nap. It is a genuine physiological intervention. And when you explain it to a three-year-old as their own special rest time rather than as a failed nap, they often accept it more readily than you expect.

The structure of quiet time matters. It should happen at roughly the same time each day, ideally in the same space. One hour is the target length, though some children do better with forty-five minutes and others stretch comfortably to an hour and a half. The space should be calm and low-stimulation: not the living room with the television, not the kitchen while you are cooking, not somewhere with high sibling traffic.

Diagram 6.3: Quiet Time as a Bridge After Nap Drop

What goes into quiet time is worth careful thought. Books are excellent. Audiobooks and music are fine. Simple puzzles or building toys can work. Screens, even educational ones, are not a good choice for quiet time because they maintain the arousal that quiet time is meant to reduce. The goal is a gentle parasympathetic shift, not entertainment.

For the parent, quiet time also serves a real recovery function. A parent who gets forty-five minutes of calm during the day is a meaningfully different parent at 5 p.m. than one who does not. You are not obligated to be productive during quiet time. Rest is productive. If you can lie down for even twenty minutes, do it. The afternoon will be better for both of you.

Some families maintain quiet time well into the early school years, even when it is clear that napping is no longer an option. A six- or seven-year-old who comes home from school and has thirty minutes of quiet time before the afternoon activities begin is often a calmer, more cooperative child for the rest of the day. The benefit does not end when the nap does.

One practical note: on days when your child skips quiet time entirely, move bedtime earlier by at least thirty minutes, sometimes forty-five. Do not wait to see how the evening goes. Move it preemptively. An overtired preschooler at 7 p.m. is much harder to settle than a tired preschooler at 6:15. The window for an easy bedtime is real, and it closes.

Nightmares: What They Are and How to Help

It is 2:30 in the morning, and your five-year-old is standing next to your bed, shaking. She is crying hard, fully awake, and she wants to tell you about what she saw. There were dogs, she says. Big ones. They were chasing her, and she could not run fast enough.

This is a nightmare. And while it is disorienting to be pulled from sleep for it, you are actually in familiar territory. She is awake. She knows you are there. She is looking for comfort, and she will find it with you.

Nightmares are REM-based events. They happen during dreaming sleep, which means they are concentrated in the second half of the night when REM pressure is highest. Your child wakes fully, or close to fully, from the nightmare. The dream content is accessible to them in the moments after waking, and it feels very real. The emotional tone is almost always fearful: being chased, falling, losing a parent, encountering something threatening.

Diagram 6.4: Nightmare vs Night Terror in the Sleep Cycle

What nightmares are not is a sign that something is psychologically wrong with your child. Nightmares are essentially universal in the preschool years, and their content maps onto the developmental preoccupations of this age. Separation, being overwhelmed, being small in a world of large things, losing safety: all of these are themes that a three-to-five-year-old's brain is actively working through during the day, and the dreaming brain recycles them at night.

Developmental research suggests that nightmare frequency often spikes during periods of transition or stress: starting preschool, the arrival of a new sibling, a move, a change in family structure. The timing is not accidental. The brain is processing change, and REM sleep is where processing happens. If your child is having more nightmares than usual, it is worth asking yourself what is new or hard in their daytime life right now.

The parent response to a nightmare is warm, calm, and brief. Go to your child or bring them to you. Hold them. Say clearly, "That was a scary dream. It is not real. You are safe. I am here." Let them tell you what happened if they want to. Do not minimize ("It was just a dream, go back to sleep"), but also do not elaborate or analyze at 2 a.m. Comfort first. Explanation later, if at all.

Returning to sleep after a nightmare can be harder for preschoolers than for older children or adults, partly because their emotional regulation is still developing and partly because the fear can persist even when the child knows, cognitively, that the dream is over. A nightlight helps. Staying in the room until the child is drowsy is reasonable. Bringing the child into your bed occasionally after a nightmare is not a habit-forming catastrophe; it is a normal human response to a frightened child.

Where parents sometimes run into trouble is in the extended processing they do at bedtime in response to recurring nightmares. Talking at length about the dream content, asking leading questions, or running through all the things that might make the nightmares stop can inadvertently amplify the child's focus on the fear. A brief, warm acknowledgment is genuinely more effective than a long reassurance marathon.

Night Terrors: A Different Animal Entirely

Here is the scenario that parents describe as the single most frightening thing they have witnessed in their child. It is 10:30 or 11:00 at night, about ninety minutes after your child went to sleep. You hear screaming, not a call-for-you kind of cry but an escalated, primal, terrified scream. You go in and find your child sitting up in bed or standing in the corner, eyes wide open, face flushed. She is screaming. She looks awake. But when you try to comfort her, she pushes you away or looks right through you as if you are not there. She does not respond to her name. She cannot be soothed. And in the morning, when you ask her about it, she has no memory of the event.

That is a night terror.

Diagram 6.5: Where Parasomnias Happen in the Sleep Cycle

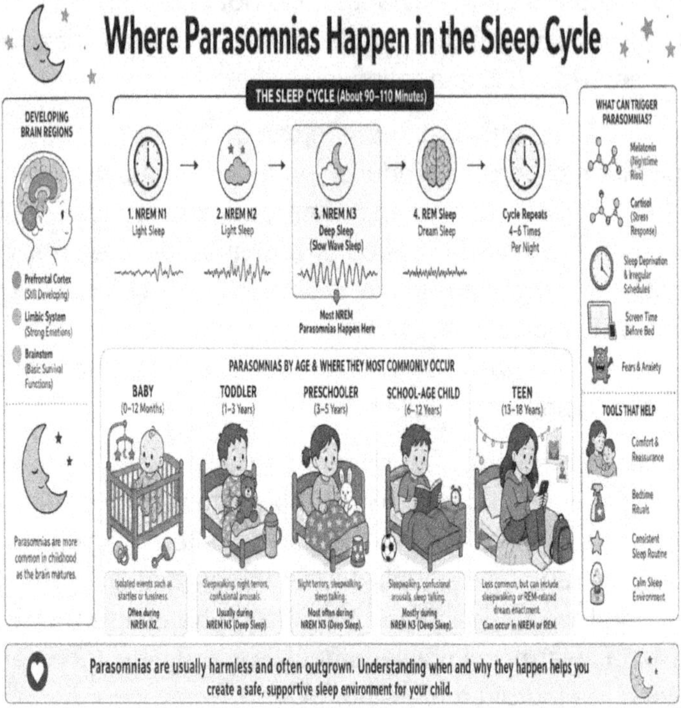

Night terrors are non-REM events. Specifically, they arise from the very deepest portion of slow-wave sleep, Stage N3, during a partial arousal. The brain is trying to transition out of the deepest sleep stage and does not complete the transition cleanly. What results is a state that looks like wakefulness from the outside but is not. The child's eyes may be open. They may move around, speak, or cry. But the cortex, the thinking, aware part of the brain, is still largely offline. This is why comfort does not work and why the child has no memory afterward.

Night terrors are most common between ages three and eight, with peak frequency in the preschool years. They tend to run in families, so if you or your partner had them as a child, your child has a higher likelihood. They are more common when a child is overtired, ill, running a fever, or under unusual stress. Sleep deprivation is the single most reliable trigger: a child who missed a nap or had a late night is significantly more likely to have a night terror that evening.

The parental response to a night terror is nearly the opposite of the response to a nightmare. You do not try to wake the child. You do not try to insert yourself into the middle of the episode to provide comfort, because the child cannot register your comfort and your attempts to intervene can actually prolong or intensify the event. What you do is stay close, keep your child physically safe (make sure they are not near stairs, sharp corners, or anything they could fall on), and wait. Most night terrors resolve within two to fifteen minutes. When the episode ends, the child will often lie back down and return to deep sleep as if nothing happened.

In the morning, do not recount the night terror to your child in front of them. They have no memory of it, and describing it in vivid detail can create secondary anxiety about something their brain did not experience. You can tell your child, briefly and matter-of-factly, that they had a restless night, but there is no need to replay the drama.

Scheduled awakenings are a well-researched strategy for frequent night terrors. If your child reliably has a night terror at a predictable time, roughly sixty to ninety minutes after falling asleep, you can gently rouse them (not fully awaken, just enough to shift their sleep stage) about fifteen to twenty minutes before the expected event. This interrupts the deep-sleep trajectory and can prevent the partial arousal from occurring. It is not always necessary, but for families dealing with nightly or near-nightly terrors, it is worth trying for a week.

Sleepwalking and Confusional Arousals

Sleepwalking belongs to the same family of events as night terrors. They are all non-REM parasomnias, events that occur during partial arousal from deep slow-wave sleep. What differs is the degree of motor involvement and the behavioral presentation.

In confusional arousals, the child appears awake, is sitting up, may be mumbling or crying, but is clearly not fully conscious. They may say strange or repetitive things. They may resist being held or directed back to bed. The episode typically lasts a few minutes and ends in continued sleep. Confusional arousals are extremely common in preschoolers and are almost always outgrown by middle childhood.

Sleepwalking involves actual ambulation. The child gets out of bed and walks around. Eyes are usually open. Movement is often purposeful-looking but disorganized. A sleepwalking child may walk to the bathroom, try to open the front door, or wander into a sibling's room. The hazard is not walking itself but the environment: stairs, exterior doors, objects they might trip over, and windows they might approach.

Diagram 6.6: Sleepwalking Safety Setup for Preschoolers

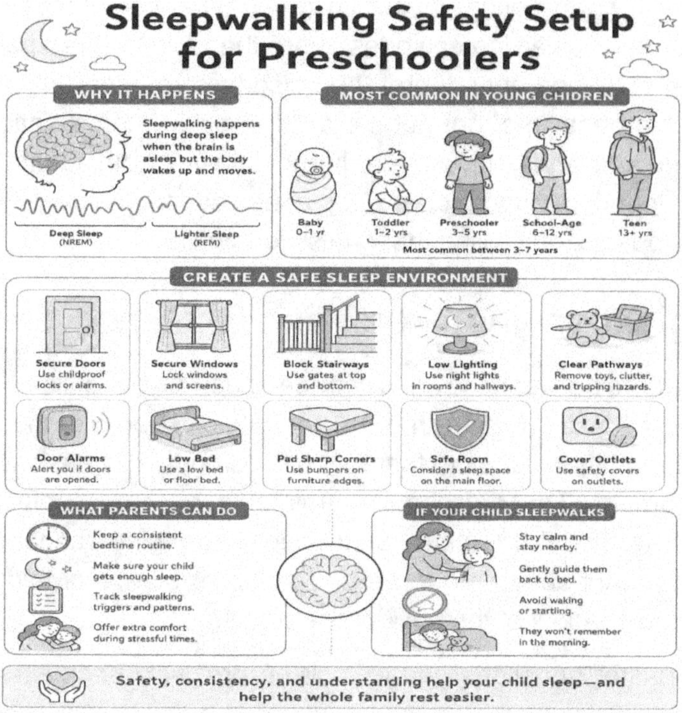

The safety setup for a sleepwalking child is straightforward. A stair gate at the top of the stairs is the most important single intervention. A door alarm or bell on the child's door alerts when the child leaves the room. Locking exterior doors at night with a lock the child cannot reach or operate is worth doing from the time you have any reason to think sleepwalking is possible. You do not need to sleep-proof the entire house, but the high-consequence hazards deserve direct attention.

Like night terrors, sleepwalking is more common when the child is overtired or has had a disruption to their normal sleep pattern. A trip, a late night, an illness, or a stretch of missed naps can all increase the frequency. The single most reliable way to reduce the frequency of all non-REM parasomnias in a preschooler is to protect total sleep time and maintain a consistent schedule.

One thing worth noting: if your child is having very frequent night terrors or sleepwalking episodes, more than two or three per week over an extended period, or if the episodes are especially intense or involve behaviors that put the child at risk, a conversation with your pediatrician is appropriate. In most cases, nothing underlying is to be found. But obstructive sleep apnea can trigger frequent non-REM arousals, and ruling that out is worthwhile.

Imagination, Monsters, and Bedtime Fears

Somewhere between the ages of three and five, your child becomes capable of a kind of fear that did not fully exist before: the fear of something that is not there. This is not regression. It is development. The preschooler brain has become sophisticated enough to model threats it cannot see, to imagine scenarios that have not happened, and to believe in the existence of things that live in the shadow at the back of the closet.

The monster under the bed is a cliché precisely because it is so universal. Across cultures and continents, preschool-age children report fear of things in the dark. The content varies. Sometimes it is monsters. Sometimes it is robbers. Sometimes it is a specific image from a television show or a story. Sometimes it is an abstract badness with no name. But the structure is consistent: darkness, aloneness, the edge of sleep, and a frightening something that might be there.

Diagram 6.7: Common Bedtime Fears by Age, Three to Eight

The developmentally appropriate response to a preschooler's fear of monsters is not to explain that monsters are not real. That response, however rational, does not land for a three- or four-year-old whose brain does not yet operate primarily in the logical register. What works better is to acknowledge the fear while gently structuring the environment to make the child feel safer.

A nightlight is one of the most effective and underrated sleep tools for this age group. Children who are afraid of the dark and have access to a dim, steady nightlight fall asleep faster and report less nighttime distress than those without access to one. The light does not need to be bright. It just needs to be enough to see the shapes of familiar things. Knowing the bookshelf is the bookshelf and not something else is genuinely calming to a preschool brain.

Comfort objects- the much-loved stuffed animal, the special blanket, the soft toy that has been sleeping in the same bed since infancy- serve a real attachment function at this age. They are transitional objects in the clinical sense: they carry the emotional residue of safety and connection, and their presence at sleep onset lowers arousal and eases the transition to sleep. If your child has a comfort object, guard it carefully. Losing it is not a minor event.

The Monster Spray Method and Other Strategies

There is a parenting tool that lives somewhere between placebo and genuine intervention, and parents who have used it love it: the monster spray. A small spray bottle filled with water, sometimes scented with a bit of lavender. Label it. Tell your child this spray keeps monsters away. Spray the closet, under the bed, and the corners of the room. Hand the bottle to your child and let them participate.

Does monster spray keep monsters away? Obviously there are no monsters to keep away. But what it does, from a behavioral and psychological standpoint, is genuinely useful. It gives the child a sense of agency over their fear. It ritualizes the transition from alertness to settled safety. It involves the parent in a concrete, actionable way that confirms the child is protected. And the lavender scent, if used, has a mild anxiolytic effect in its own right. The fear is real. The spray addresses the fear. That is not magical thinking on your part; it is good parenting.

Diagram 6.8: The Monster Spray Method, Step by Step

A few additional bedtime-fear strategies are worth knowing. First, avoid confirming that something frightening actually exists, even while using tools like monster spray. "The spray keeps monsters away" is different from "Yes, there was a monster there, but now it is gone." The former addresses the fear without ratifying it. The latter can amplify it.

Second, timing matters for fear conversations. Brief, warm acknowledgment at bedtime is appropriate. Extended processing ("Tell me more about what the monster looked like, what does it do, how big is it") does not belong at 8 p.m. It ramps up the very mental activity you are trying to quiet. Daytime, with a snack, good lighting, and a relaxed, conversational tone, is the right time for that kind of talk if your child wants it.

Third, keep the check-in ritual short and consistent. Many families find that offering one brief check-in after the child is in bed, a one-minute visit after five minutes, for example, followed by progressively longer gaps, gives the child a predictable anchor point and reduces the anxiety of wondering when someone will come. If the child knows you are coming back in five minutes, they have something to focus on besides the dark.

Fourth, do not dismiss the fear, even if you are tired, the bedtime is already running 40 minutes late, and you have an early morning tomorrow. "There is nothing there, go to sleep" is technically accurate and practically counterproductive. "I hear you, I know that felt scary, you are safe, and I am right outside" takes about the same number of seconds to say and produces a very different outcome.

Starting Preschool and Schedule Disruptions

There are few bigger changes in a young child's daily life than starting a group care program for the first time, whether that is a preschool, a daycare center, a pre-K program, or any other setting where your child is suddenly spending six or eight hours a day in a new environment with new adults and unfamiliar peers.

The sleep impact of this transition is real and almost universal. The first weeks of a new school environment commonly produce one or more of the following: difficulty falling asleep at night, more frequent waking, earlier morning waking, increased bedtime resistance, a return of separation anxiety at drop-off that bleeds into night waking, and changes to the nap (either refusing the nap at school, or napping longer than usual and then being unable to fall asleep at the normal bedtime).

Diagram 6.9: Preschool Transition Sleep Plan

None of this means you made the wrong choice about school. It means your child is processing an enormous amount of new information, social input, and sensory stimulation every day, and their nervous system is working hard to metabolize it. Sleep is where that work happens, and it can temporarily make sleep itself more difficult.

There are things you can do to support the transition. The most important is to protect the home sleep environment as a place of predictability and calm. Whatever the school schedule does to nap timing, keep the bedtime routine itself absolutely consistent. Same order of events, same ending, same warm send-off. The routine is the signal that the intense, stimulating part of the day is over and the safe, familiar part has begun.

If your child naps at school, find out the school's nap schedule and ask when naps typically end. A child who naps until 3:30 at school is not going to be ready for a 7:30 bedtime. Shifting bedtime to 8:15 or 8:30 during the school week, and keeping a 6:30 or 7:00 bedtime on weekends when nap timing is in your hands, is a reasonable adjustment.

Watch for the overtiredness pattern that can build across the school week. Many preschoolers are fairly functional Monday and Tuesday, start showing cracks on Wednesday, and are truly running on empty by Friday. If this is your child, Friday-evening plans that require social performance or impulse control will be hard. Friday afternoons are a good time for screen-free quiet time, early dinner, and an early bedtime, not the afternoon for a birthday party or a long grocery trip.

Separation anxiety, which is developmentally typical in this age group and can intensify around major transitions, often shows up at bedtime as the child suddenly needs one more hug, one more check, one more assurance that you will be there in the morning. Handle this with warmth and brevity. One reassurance, confidently given, is more effective than ten reassurances given with escalating anxiety. "I will be here in the morning. I love you. Goodnight." And then you leave.

Two-parent (or two-caregiver) consistency deserves a direct mention here. Preschoolers are developmentally wired to find inconsistency and use it. This is not manipulation in an adult sense. It is normal human learning: if the rule bends with one person, the child will keep trying to bend it. If both adults in the household have different bedtime rules, the child will negotiate endlessly for the more permissive version. Getting aligned in advance, on bedtime, on how many times you go back in, on whether books after lights-out is allowed, on what happens when the child appears in your doorway at midnight, saves enormous nightly friction. Writing it down somewhere both adults can see it is not overkill. It is a practical tool.

Try This Tonight

For this age range: 3 to 5 years

1. Tonight, run a four-step bedtime routine: wash up or bath, pajamas, two stories, lights out. Do not add anything else. If your child asks for more, name the routine clearly: "We did our four things. That is our bedtime. I love you, goodnight." Keep it the same tomorrow night.

2. If your child expresses a fear (monster, dark, something scary), do one brief acknowledgment and one concrete action. Say "I hear you, you are safe," then choose one tool: turn on the nightlight, do a quick monster-spray of the room, or hand them their comfort object. Then say goodnight and leave. Do not extend the conversation.

3. If your child had a nightmare recently and is anxious about sleep, try briefly telling them a calm, short story as the last part of your routine, one where the child in the story is safe, loved, and falls asleep peacefully. End the story before your child is fully asleep so they make the final transition on their own.

4. If you know your child tends to have night terrors, check the bedtime. Overtiredness is the primary trigger. Move tonight's bedtime thirty minutes earlier than usual. Protect total sleep time this week.

5. If your child dropped the nap recently, move bedtime earlier by at least thirty minutes starting tonight. Do not wait to see how the evening goes. Do it proactively. An overtired preschooler is harder to settle than a tired one.

6. Review the sleep environment for the safety basics: nightlight on, door cracked if your child prefers it, comfort object in the bed, stair gate secured if sleepwalking has ever occurred. These are not dramatic interventions. They are the infrastructure that makes everything else easier.

7. If you share bedtime duties with another caregiver, take five minutes today, not at bedtime, to align on two specific things: what the routine looks like and what you both do when the child comes out of their room after lights-out. One agreed response is worth more than ten improvised ones.

Parent Takeaways

- Most children drop the nap between ages three and five. Quiet time replaces it, and bedtime should move earlier to compensate for the lost daytime sleep.

- Nightmares and night terrors are two different neurological events. Nightmares occur during REM sleep, often late at night, and the child is fully awake and remembers them; your job is to comfort and reassure. Night terrors come from deep non-REM sleep, occur early in the night, and the child is not conscious and will not remember; your job is to stay calm, keep them safe, and wait.

- The imagination that powers your preschooler's creativity also powers their bedtime fears. A brief, warm acknowledgment plus one concrete tool (nightlight, monster spray, comfort object) is more effective than extended reassurance or a logical argument.

- Starting preschool disrupts sleep for almost every child, usually temporarily. Protect the home routine as a source of consistency during the adjustment, and watch for signs of increasing overtiredness throughout the school week.

- Overtiredness is the number one trigger for night terrors, sleepwalking, and increased nightmare frequency. When sleep events escalate, the first question is always whether your child is getting enough total sleep.

7 School-Aged Sleep: Ages 6 to 11

Your eight-year-old went to bed at nine. He was asleep by nine-thirty. He woke up at seven. By any surface measure, that sounds like a child who is sleeping. So when his teacher calls to say he has been distracted in class, irritable at recess, and struggling to hold on to the reading material he covered last week, your first thought is probably school stress, a learning challenge, or maybe just a rough patch. Sleep is probably not even on the list.

That is exactly the trap this chapter is here to help you avoid.

The school-age years, roughly ages 6 to 11, have a reputation for being the "easy" years of parenting. The newborn chaos is long gone. The toddler tornado has calmed. Your child dresses herself, makes her own breakfast, and can spend twenty minutes in a room alone without catastrophe. Sleep-wise, things feel manageable. Bedtime is not the hourlong battle it was at three. Most kids this age go down without much drama.

But "not much drama at bedtime" and "getting enough sleep" are two very different things. And the gap between those two realities is where many school-age struggles quietly live.

A child who is chronically short on sleep does not look sleepy. She looks restless. She looks reactive. She looks like the kid who cannot sit still in the classroom, who melts down over minor frustrations at home, who fights with friends over small misunderstandings that a rested child would shrug off entirely. She looks to many adults like a child with a behavior problem. Sometimes she looks like a child with ADHD. In many cases, she is simply a child who is not getting enough sleep.

This chapter is going to walk you through all of it: what the science says about how much sleep your school-age child actually needs (spoiler: it is probably more than she is getting), why a sleep-deprived kid looks wired instead of tired, how homework loads and screen habits and over-packed activity schedules are quietly eating away at the sleep window, and what a realistic school-night routine looks like for this age group. There is a lot here that will feel immediately recognizable, and probably a little relieving once you see it laid out.

You are not imagining it. And it is not just your kid. This is happening across classrooms and households at a scale that most parents do not quite realize until someone puts the numbers in front of them. So let us start with the numbers.

Chapter Summary

This chapter covers the sleep needs, sleep habits, and sleep challenges of children ages six to eleven. These are sometimes called the quieter years for sleep. Still, they are more accurately the under-monitored years: children in this window often lose significant sleep, presenting as behavioral or academic problems rather than sleep problems. The American Academy of Pediatrics recommends 9 to 12 hours of sleep for this age group, and most American school-age children consistently fall short of that range.

You will learn why a tired child this age looks hyperactive rather than sleepy, why that pattern gets confused with ADHD, and what the research says about how even moderate sleep shortfalls affect academic performance, emotional regulation, and social functioning. You will also find practical tools for evaluating your own child's sleep, navigating the homework and screen pressures that eat into the weeknight sleep window, managing the over-scheduling trap, and building a school-night routine that actually holds up in the real world. The habits built in this window matter enormously: the teen years bring a dramatic biological shift in sleep timing, and children who enter adolescence with strong sleep foundations handle that shift far better than those who do not.

What This Chapter Will Cover

- Sleep need at ages 6 to 11. The AAP recommends 9 to 12 hours of sleep for children in this age range, and the evidence supporting that recommendation is substantial. Most American children in this age group are sleeping between eight and nine hours on school nights, which means many are running a meaningful deficit every single week. This section helps you figure out exactly where your child stands and what the gap looks like in real terms.

- Why sleep-deprived kids look hyperactive, not sleepy. The paradoxical arousal response is one of the most important things to understand about pediatric sleep science. When a child's brain is underslept, it floods with stimulating neurochemicals in an attempt to compensate, producing restlessness, impulsivity, and emotional volatility rather than the drowsiness you might expect. This section explains the neuroscience in plain language and helps you recognize the pattern.

- Sleep deprivation versus ADHD. Because chronic sleep loss in children mimics the behavioral profile of attention-deficit/hyperactivity disorder, it is often misdiagnosed or missed entirely. This section explains the overlap, the key differences, and why a sleep evaluation should be part of any ADHD workup.

- How slow-wave sleep builds academic skills. The deepest stage of sleep is when memory consolidation occurs, when the skills your child practiced and the information she absorbed are transferred into long-term storage. This section explains why cutting sleep short also shortens learning.

- The homework, screen, and bedtime collision. On a typical school night, a child might have homework until eight, then an hour of screen time to decompress, and not actually be in bed until nine-thirty or ten. This section maps that trap and offers realistic strategies for getting ahead of it.

• Over-scheduling and sleep debt. Practices that run until eight, music lessons on weeknights, travel sports on weekends: the modern childhood schedule is genuinely at war with adequate sleep. This section helps you think through the tradeoffs without guilt or judgment.

• Social jet lag and weekend sleep debt. Sleeping in on Saturday and Sunday to "catch up" shifts the internal clock in ways that make Monday morning feel like flying across time zones. This section explains what that does to your child's brain and body, and how to manage the weekend transition without losing all flexibility.

• The school-night routine that actually works. Consistent timing, light management, and a wind-down sequence that the whole family can sustain: this section gives you a concrete, step-by-step framework built around what the research actually supports.

• Sleep and emotional regulation. Sleep is not just about focus and memory. It is the overnight repair system for your child's emotional brain. A well-rested child handles conflict better, recovers from disappointment faster, and experiences anxiety at a lower intensity. This section connects the sleep dots to the school-age child's emotional life.

How Much Sleep Does Your Child Actually Need?

Let us start with the baseline, because it matters more than most parents realize.

The American Academy of Pediatrics recommends that children between the ages of six and twelve consistently sleep nine to twelve hours per night. That recommendation is not arbitrary. It is grounded in decades of research connecting sleep duration to cognitive development, immune function, metabolic health, emotional regulation, and behavioral outcomes in this age group. Nine to twelve hours. Every night.

Now consider what is actually happening. Large-scale population surveys consistently find that American school-age children are sleeping, on average, between eight and nine hours on school nights. That might not sound like a crisis, but when you do the math, a child who needs 10 hours and gets 8.5 is running a 90-minute deficit every single weekday. By Friday, that child has accumulated roughly 7.5 hours of sleep debt. That is not a small shortfall. That is the equivalent of pulling two all-nighters across the course of a school week, stretched out in smaller doses so the impact is less obvious but no less real.

Diagram 7.1: School-Age Sleep Need vs Reality

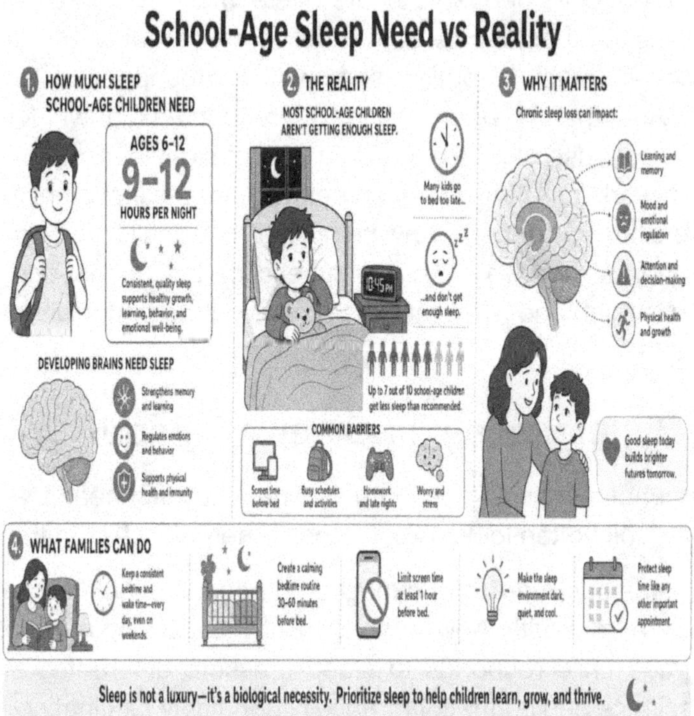

What makes this particularly tricky is that children do not show this deficit to you the way adults do. Chronically sleep-deprived adults feel tired. They yawn, they drag, they reach for coffee, they know something is off. Children, especially those in the six-to-eleven age range, have nervous systems that do not respond to sleep deprivation in the same way. Instead of shutting down, it revs up. The brain, sensing that it is underfueled, reaches for any available stimulant it can manufacture. The result is a child who appears wired, wound up, and difficult to settle, rather than one who looks like she needs sleep.

Part of the reason for the chronic shortfall is simply arithmetic. Consider what a typical school morning looks like: school buses arrive, or the family car has to leave, somewhere between seven and eight in the morning for most American elementary-age children. If a child needs 10 hours of sleep and has to be up at 7, she needs to be asleep by 9. In practice, that means she needs to start her wind-down routine somewhere around eight or eight-fifteen. And that is where things start to break down for most families, because eight o'clock on a school night is also prime time for homework, for after-dinner screen use, for the last bit of outdoor play, and for the general unwinding that both children and adults need after a long day. The arithmetic of the school-age sleep window is tight, and most families are losing the battle without fully realizing it.

The Hyperactive-Not-Sleepy Paradox

Here is what surprises most parents when they first hear it: a sleep-deprived child often looks like the most energetic child in the room.

This seems backward. We expect tired people to look tired. And in adults, that is mostly how it works. But the developing brain of a school-age child responds to sleep pressure differently, and the neuroscience behind that difference is genuinely fascinating. When a child is underslept, the brain does not simply slow down. Instead, it activates what researchers call a compensatory arousal response: a flood of stimulating neurochemicals, including dopamine and norepinephrine, that serve as an emergency alert system. The brain is essentially screaming at itself to stay functional despite the deficit.

The result is paradoxical arousal. The child does not look tired. She looks bouncy, silly, reactive, impulsive, and difficult to redirect. She might be laughing too loudly at nothing, picking fights with her sibling for no real reason, unable to sit still for a meal, or losing it completely over something that would barely register on a normal day. To the adults in the room, this looks like a behavior problem. To the child, it feels like a nervous system that will not quiet down, and she has no vocabulary to explain why.

Diagram 7.2: The Hyperactive-Not-Sleepy Paradox

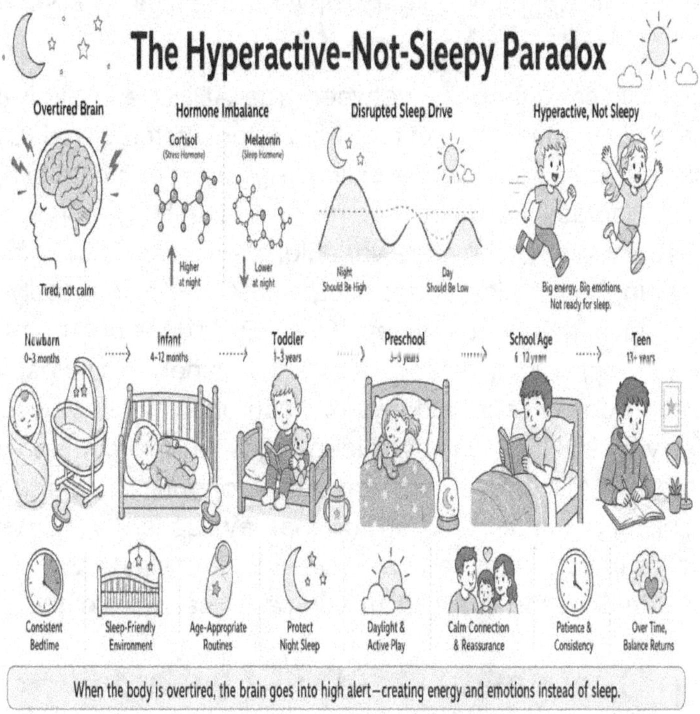

This matters enormously for how parents, teachers, and pediatricians interpret what they are seeing. When a child presents with hyperactivity, inattention, and emotional dysregulation, the first instinct in most clinical settings is to consider ADHD. And ADHD is absolutely real, important to identify, and deserving of proper support when it is present. But sleep deprivation and ADHD share a strikingly similar surface presentation, and the research suggests that sleep problems are both underidentified and undertreated in children who have already been labeled with attention difficulties.

Studies tracking large groups of school-age children have found that improving sleep duration and quality produces measurable reductions in teacher-reported ADHD-like symptoms, even in children who were not formally diagnosed. That is remarkable. It suggests that for a meaningful proportion of children showing these symptoms, sleep is a primary driver rather than a secondary concern.

The key clinical difference between true ADHD and sleep-driven hyperactivity comes down to a few important factors. True ADHD is pervasive across settings: the child struggles at home, at school, at birthday parties, in quiet one-on-one situations. Sleep-driven hyperactivity tends to be more variable, often worse in the afternoon and evening when sleep debt accumulates, and noticeably better on mornings after a longer sleep. True ADHD does not dramatically improve when you extend sleep by ninety minutes; sleep deprivation does. That said, the two can absolutely co-occur, and children with ADHD often have sleep difficulties that compound their symptoms. If ADHD is being considered for your child, a thorough sleep history should be part of that evaluation. Any pediatric assessment that does not ask detailed questions about sleep timing, duration, and quality is missing a critical piece of the picture.

Sleep Deprivation vs ADHD: What to Know

It is worth spending a little more time here, because this is an area where many families feel confused and sometimes unheard.

If you have sat in a pediatrician's office or a school meeting and heard concerns about your child's attention or behavior, and you have been thinking quietly in the back of your mind that he seems tired, you may well be right. The overlap between sleep deprivation and ADHD is not a fringe theory. It is a documented clinical reality, and it has been studied in enough depth that major pediatric sleep organizations now recommend sleep evaluation as a first step in any ADHD workup.

Diagram 7.3: Sleep Deprivation vs ADHD Symptom Overlap

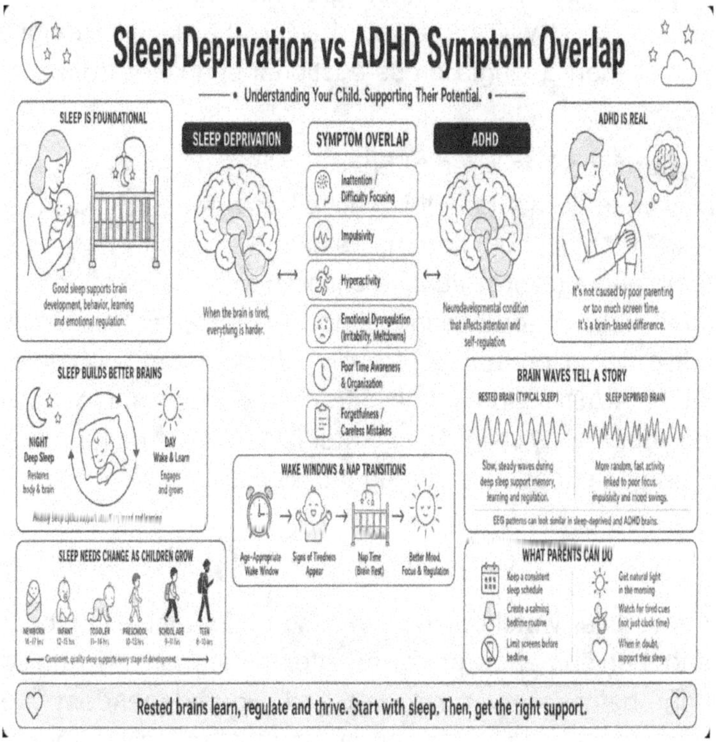

What chronic sleep deprivation looks like in a school-age child: difficulty staying on task during longer activities, impulsive responses to frustration, trouble transitioning between activities, emotional sensitivity that seems disproportionate, and restlessness that teachers often describe as "can't sit still." These are also, word for word, many of the symptoms on a standard ADHD screening checklist. The overlap is not coincidental. Both conditions affect the same prefrontal circuits, the same dopamine pathways, the same executive function networks.

What helps you tell them apart at home: track your child's behavior in relationship to how much she slept the night before. Keep a simple log for two weeks. Note bedtime, approximate sleep-onset time, wake time, and any notable behaviors at school or home. If the pattern shows a clear correlation between sleep amount and behavioral quality, that is meaningful information. Take it to your pediatrician. Most will appreciate having that kind of longitudinal data rather than a single cross-sectional snapshot from a school referral.

And if your child has already been evaluated and diagnosed with ADHD, please do not take this section as a reason to dismiss that diagnosis. What it should prompt is a conversation with your child's care team about sleep, because children with ADHD often have co-occurring sleep disorders including insomnia, restless legs, and sleep-disordered breathing. Treating those sleep problems does not eliminate ADHD, but it consistently reduces symptom severity. Helping your ADHD child sleep better is one of the highest-leverage ways you can support her during the day.

Slow-Wave Sleep and Memory Consolidation

If you have ever watched your child study for a spelling test, go to bed, and then somehow perform better in the morning than she did the night before, you have watched sleep-dependent memory consolidation in action. It is one of the most reliably demonstrated phenomena in sleep science, and it has direct implications for every homework assignment, every practice session, and every classroom lesson your school-age child is absorbing.

Here is what is happening. During the deepest stage of non-REM sleep, called slow-wave sleep or deep sleep, the brain does something extraordinary. It replays the day's learning. The hippocampus, which acts as a temporary holding area for new information, transfers its contents to the cortex for long-term storage. The neural circuits activated during the day's practice are strengthened and organized. Skills that were effortful become more automatic. Fuzzy concepts become clearer.

Diagram 7.4: Slow-Wave Sleep and Memory Consolidation

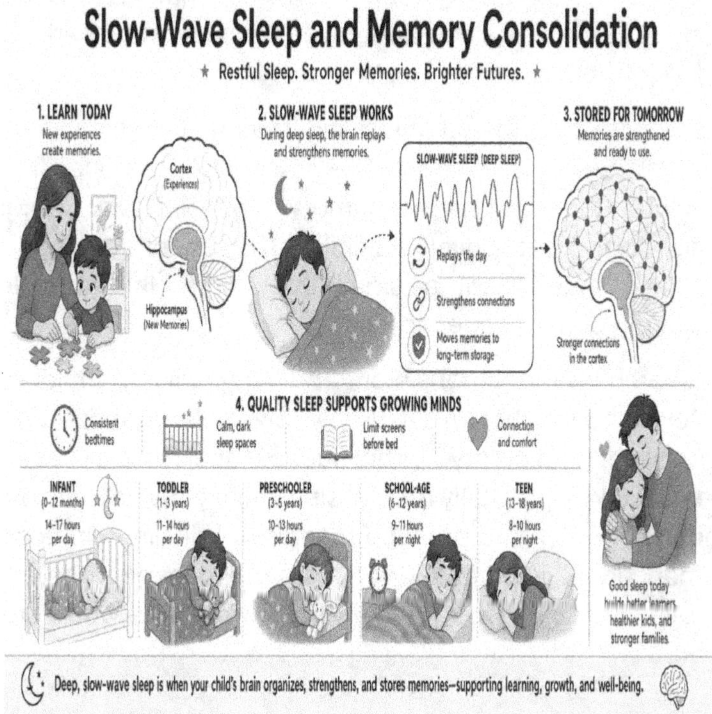

This process is not a passive recording. It is an active, selective process that prioritizes information the brain has tagged as important, makes connections between new learning and existing knowledge, and prunes out the noise. During slow-wave sleep, the brain is doing the intellectual equivalent of organizing a messy desk: sorting, filing, and discarding. And it cannot do that work if sleep is cut short.

Here is the critical piece for school-age children: slow-wave sleep is concentrated in the first half of the night. The early hours of sleep, roughly the first four to five hours, are dominated by deep slow-wave cycles. REM sleep, which supports emotional processing and creative thinking, concentrates in the second half of the night. When a child goes to bed late and wakes up early for school, she does not lose sleep uniformly from both halves. She compresses or eliminates the second-half REM sleep almost entirely while protecting the slow-wave sleep in the first half. But when she regularly accumulates sleep debt, even the slow-wave periods start shrinking.

A child who is going to bed at ten and waking at six-thirty is not just short on sleep in the abstract. She is short on the specific sleep architecture that supports memory consolidation, emotional regulation, and the cognitive flexibility that school demands every single day. That is not a small thing. That is the biological underpinning of her academic life.

The research on this is consistent and fairly striking. Children who sleep fewer hours show measurably lower performance on tasks involving working memory, verbal fluency, and complex reasoning. Studies that experimentally restrict sleep in healthy children by even one hour for five nights produce meaningful declines in academic performance that parents and teachers can observe. The effect is not subtle. And the good news is that it reverses when sleep is restored.

The Homework-Screen-Bedtime Trap

Let us talk about what actually happens in most homes on a school night, because the pattern is remarkably consistent.

School ends somewhere between two-thirty and three-thirty. The child has some afternoon activity, or comes home and decompresses, or does both. By five o'clock, most families are thinking about dinner. Dinner runs until six or six-thirty. Then comes the homework. Depending on the grade, homework might take 30 minutes or 2 hours. Most research suggests that homework for elementary school children should be limited to about 10 minutes per grade level per night (so 30 minutes for a third-grader). Still, in many schools and households the reality is considerably more. Heavy homework nights can push the start of any wind-down time to eight-thirty or nine.

Diagram 7.5: The Homework-Screen-Bedtime Trap

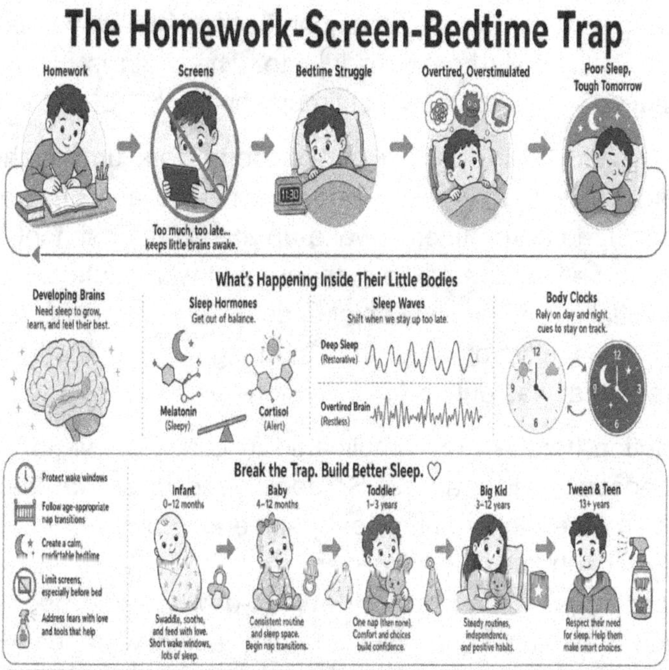

After homework, most children need some transition time before they can sleep. They are cognitively activated from schoolwork. They are often frustrated or stressed from whatever challenges came up. And this is the moment when screens tend to enter the picture. A child who just spent ninety minutes on math worksheets reasonably wants to watch something or play something to decompress. That impulse is completely understandable. The problem is that screens, particularly interactive ones like video games, extend the activation window rather than closing it.

The blue light emitted by screens suppresses melatonin production, which is the signal the brain uses to begin its transition toward sleep. But that's not actually the most important screen problem at this age. The more significant issue is cognitive and emotional arousal. An exciting video game, a funny YouTube video, a group chat with friends: all of these engage the reward system, the social processing circuits, and the attention networks in ways that make sleep physiologically harder to initiate. The brain is not ready to sleep because it is still chasing stimulation.

So the typical chain looks like this: homework until eight-thirty, screens until nine-thirty, lying in bed activated and unable to fall asleep until ten or ten-fifteen, wake-up at seven for school. That is eight and a half to nine hours in theory, but with a late sleep onset and potential overnight awakenings, actual sleep time is often less. And the child wakes on a Monday morning carrying the full load of a weekend that was only partially restorative.

The solution here is not complicated in concept, even if it takes some negotiation in practice. Screens need to end at least sixty minutes before lights-out, ideally ninety. Homework needs a dedicated, bounded window that ends by eight at the latest for most elementary-age children. And the wind-down period needs to be protected the way your child's health actually depends on it, because it does. If homework is regularly running past eight, that is a conversation worth having with a teacher. Most educators, when they understand the sleep science, are genuinely receptive to that kind of feedback.

Over-Scheduling and Sleep Debt

There is a version of the modern childhood schedule that is genuinely exhausting to look at from the outside, and many parents living it are so used to the pace that they have lost track of how little unstructured time their children have. Not just unstructured daytime, but unstructured nighttime. Sleep time.

Soccer practice on Tuesday and Thursday evenings until seven-thirty. Piano on Wednesday. A travel tournament most weekends from September through November. A second sport starting in January. A language class on Friday afternoons. Birthday parties on Saturdays. And homework every night on top of all of it.

Diagram 7.6: Over-Scheduling and Sleep Debt

Each one of these activities, taken individually, is reasonable and often wonderful. Children benefit from sports, music, and social engagement. The problem is cumulative and arithmetic. When practices run until seven-thirty or eight, dinner happens at eight-thirty, homework starts at nine, and bedtime is not until ten or later, the child is sleeping seven or eight hours on a night when she needs ten. Do that five nights a week, and the weekly sleep debt is enormous.

The research on this is not ambiguous. Children who participate in organized activities to the point of sleep restriction show measurable performance declines in school, elevated cortisol levels (the stress hormone), and higher rates of reported anxiety and emotional difficulty. The activities intended to enrich their lives are, through the mechanism of sleep deprivation, actively undermining their well-being.

This is not about guilt. Every parent making these scheduling choices does so out of love and a genuine desire to give their child opportunities. But it is worth pausing regularly to ask: if I add up all the commitments on this calendar, what time does my child actually get into bed on a school night? If the honest answer is nine-thirty or later, something needs to change. That might mean dropping an activity, moving a practice to a weekend slot, or simply having a direct conversation with a coach or program director about late scheduling and its impact on children's health.

You are allowed to advocate for your child's sleep. You are allowed to say that an eight o'clock Tuesday practice is not compatible with your child getting enough sleep, and that you are going to have to make a different choice. Other parents in the same program are probably quietly feeling the same thing and waiting for someone to say it first.

Social Jet Lag in School-Age Kids

Friday arrives. Your child is visibly exhausted, probably slightly more irritable than usual, and ready to collapse after school. You let her stay up a little later on Friday night. She sleeps in Saturday morning. She stays up later Saturday night and sleeps in Sunday. By Sunday afternoon, she seems great. But Monday morning? Monday morning is a disaster.

This pattern has a name: social jet lag. And it is one of the most underappreciated sleep problems among school-age children.

Diagram 7.7: Social Jet Lag in School-Age Children

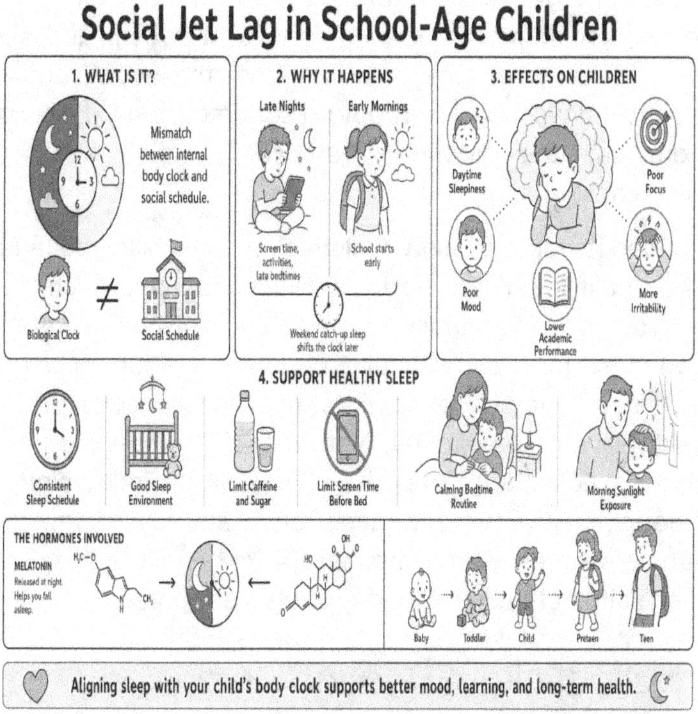

Social jet lag occurs when a child's sleep schedule shifts meaningfully between the weekend and the school week. If your child falls asleep at nine-thirty and wakes at seven on weekdays, then stays up until eleven and wakes at nine on weekends, her internal clock experiences something functionally similar to crossing two time zones every Friday and then crossing back every Sunday night. The circadian clock, which regulates when the body and brain are ready for sleep and wakefulness, does not reset overnight. It shifts gradually, at roughly one hour per day in the forward direction (staying up later) and even more slowly in the reverse direction (going back earlier).

When your child sleeps in by two hours on Saturday and Sunday mornings, her body clock shifts by two hours. Then, Sunday night, when she needs to be asleep by nine-thirty to function Monday morning, her body clock says it is seven-thirty. She is not tired. She lies in bed, she gets frustrated, and she eventually falls asleep at eleven or later. Monday morning arrives at seven, and she has gotten eight hours of sleep starting from the wrong end of her biological window. She is groggy, disoriented, and dysregulated. Teachers call this "Monday behavior," and it has a clear physiological cause.

The solution is not to eliminate all weekend flexibility, because that is neither realistic nor necessary. A reasonable approach is to limit weekend sleep timing variation to thirty to forty-five minutes in either direction from the school-week schedule. If your child typically wakes at seven on school days, let her sleep until seven-thirty or seven-forty-five on weekends. Not nine. If she needs some extra rest on the weekend, go to bed a bit earlier on Friday night rather than sleeping dramatically later on Saturday morning. This approach preserves most of the restorative benefit while minimizing the clock shift that makes Monday such a struggle.

The School-Night Routine That Works

Here is where everything comes together, and where parents often breathe a small sigh of relief, because despite all the complexity covered in this chapter, the practical solution has a fairly simple shape.

The foundation of good sleep in the six-to-eleven age range is a consistent, predictable, screen-free wind-down routine that begins early enough to allow for adequate sleep duration before the required wake time. That sentence contains every principle that matters. Let us unpack each piece.

Consistent and predictable means the same sequence of events at roughly the same time every school night. The child's brain begins to associate the sequence with the approach of sleep, and cortisol starts to drop, melatonin starts to rise, and the body begins to cool (a key physiological prerequisite for sleep onset) before the child is even in bed. Routine is doing biological work, not just logistical work.

Diagram 7.8: The School-Night Routine That Works

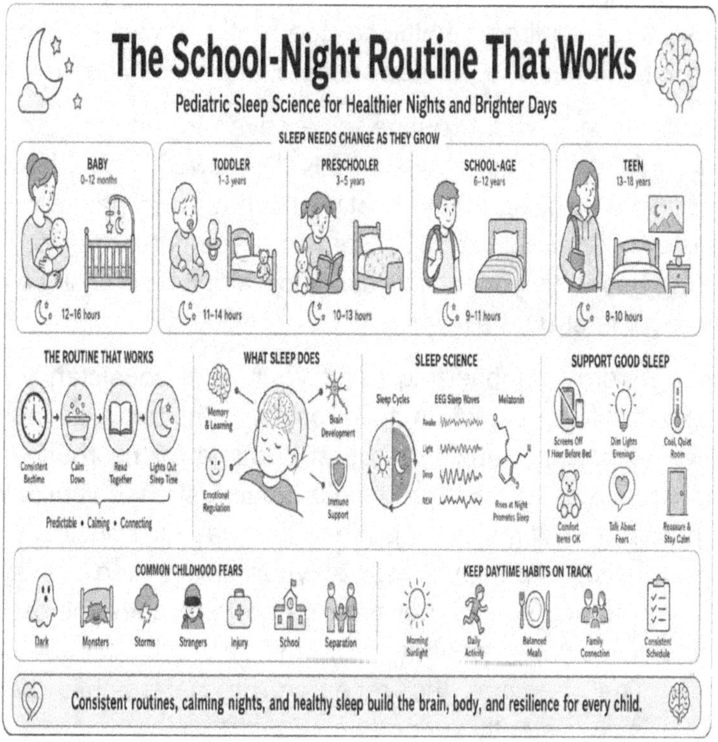

Screen-free means that all devices, including tablets, phones, gaming consoles, and, yes, television if it is in the bedroom, are turned off and put away at least 60 minutes before lights-out. This is not about punishing screen use. It is about giving the nervous system the transition time it needs. For most school-age children, sixty minutes is the minimum; ninety is better. The last hour before bed should involve calming, low-stimulation activities: a bath or shower (which triggers the temperature drop that helps initiate sleep), reading a physical book, drawing, a quiet conversation with a parent, or simple relaxation practices.

Early enough means working backward from the required wake time. If wake time is seven, and your child needs ten hours of sleep, she needs to be asleep by nine. That means she needs to be in bed and lights out by eight-forty-five at the latest, so the wind-down routine needs to start by seven-forty-five or eight. For many families, this requires rethinking the entire evening schedule, starting with dinner time, homework time, and activity commitments.

A workable school-night routine for a seven-to-ten-year-old might look like this: dinner done by six-thirty, homework with a hard stop at seven-forty-five, screens off at eight, bath and pajamas, thirty minutes of quiet reading in bed, lights out at eight-forty-five. Adjust the timing based on your child's specific wake time and sleep need, but protect the structure.

One thing worth emphasizing here: your calm consistency is the most powerful ingredient in this routine. Children this age are sensitive to parental anxiety around bedtime. If the routine feels rushed and stressed, it will produce a rushed, stressed child. When you approach bedtime as a protected, pleasant transition rather than a logistics problem to solve, your child's nervous system follows your lead. You do not need to perform cheerfulness. You need to slow down, lower your voice, and be present for about thirty minutes. That is what the research on parent-child bedtime interactions consistently supports.

Sleep and Emotional Regulation

It would be easy to read this chapter as mainly about academic performance, because much of the conversation around school-age sleep focuses on grades and attention. But there is another dimension that matters just as much, if not more: how sleep shapes the emotional life of the school-age child.

The prefrontal cortex, the part of the brain that handles impulse control, perspective-taking, conflict resolution, and emotional moderation, is still very much under construction in a six-to-eleven-year-old. It will not reach full maturity until the mid-twenties. Sleep is one of the primary inputs that allows this developing system to function at its best during the day. When sleep is adequate, the prefrontal cortex can do its job: it can put the brakes on a reactive impulse, find words for a frustrating feeling, and recognize that a misunderstanding with a friend is not a catastrophe. When sleep is insufficient, the prefrontal cortex goes offline in partial ways, and the more primitive emotional circuits take over.

Diagram 7.9: Sleep and Emotional Regulation in School-Age Children

This is why a sleep-deprived child can erupt over something that seems trivially small: a sibling who looked at her wrong, a sandwich cut the wrong way, a homework assignment with an unexpected instruction. Her emotional brain is reacting, and the regulatory system that would normally soften that reaction is not fully available. It is not a character flaw. It is a neuroscience problem.

The research on sleep and emotional regulation in children is remarkably consistent across labs and populations. Children who sleep less show lower frustration tolerance, higher rates of anxiety symptoms, poorer peer relationship quality, and lower self-reported wellbeing than age-matched peers who sleep adequately. Longitudinal studies following children over multiple years find that chronic sleep shortfalls at this age predict higher rates of anxiety and mood difficulties in adolescence, even after controlling for other risk factors. Sleep is not just a support for emotional health. It is a foundational input.

For children who are already prone to anxiety or who are going through a particularly difficult social period, this connection is especially important to attend to. Anxiety and sleep deprivation feed each other in a well-documented bidirectional loop: anxiety makes it harder to fall asleep and stay asleep, and sleep deprivation amplifies anxiety. For a child who is worried about school, friendships, or family stress, getting adequate sleep is one of the most powerful tools available. It does not eliminate the worry, but it reduces the emotional volume at which it is experienced, making it significantly more manageable.

Friendships at this age, which feel enormously high-stakes to children even when they appear small to adults, are navigated much more successfully by rested children. A well-slept child can pick up social cues more accurately, recover from awkward interactions more easily, and tolerate the ordinary friction of peer relationships without catastrophizing. These are the social skills that carry forward into adolescence, and they are genuinely trainable through adequate sleep.

None of this means that emotional difficulties in your school-age child are purely a sleep problem, or that sleep is the only lever you have. But it does mean that before assuming a behavioral or emotional challenge requires a clinical response, it is worth asking honestly whether your child is getting enough sleep. Often, addressing the sleep straightens out a surprising amount of the rest.

Try This Tonight

For this age range: 6 to 11 years

1. Calculate your child's ideal bedtime tonight. Take your child's required wake-up time, subtract ten hours (or the amount your child specifically needs based on observation), and set that as your target lights-out time. Work backward by 30 to 45 minutes to find when the wind-down routine needs to start.

2. Move all screens out of your child's bedroom right now. Phones, tablets, handheld gaming devices, and laptops should be charged overnight in a common area, such as the kitchen or hallway. If there is a television in your child's room, turn it off and consider whether it belongs there.

3. Set a screen curfew for tonight. Whatever your target lights-out time is, screens should be off sixty to ninety minutes before that. Announce this as a new family policy, not a punishment. Keep your tone calm and matter-of-fact.

4. Replace screen time before bed with one quiet activity your child enjoys. A physical book, drawing, a simple puzzle, or a quiet conversation with you. Let her have some input in choosing it so there is buy-in.

5. Check whether homework is ending early enough. If homework regularly runs past eight o'clock for an elementary-age child, plan to talk with your child's teacher this week about workload timing and sleep needs.

6. Notice your child's behavior tomorrow and connect it to sleep. Not to your child, but in your own mind. Jot down roughly how long she slept and how her morning looked. This awareness is the beginning of understanding her personal sleep pattern.

7. On Saturday morning, try to keep her wake time within forty-five minutes of her school-week wake time. Notice whether Sunday night bedtime is easier or harder when you limit the morning shift.

Parent Takeaways

- The AAP recommends nine to twelve hours of sleep for children ages six to twelve, and most American children in this range fall significantly short on school nights.
- A sleep-deprived school-age child often looks hyperactive, impulsive, and emotionally reactive rather than sleepy, which means sleep deprivation is frequently mistaken for behavioral or attention problems.
- Because chronic sleep loss mimics ADHD symptoms closely, any evaluation for attention difficulties should include a thorough sleep history and, when possible, a trial of improved sleep before or alongside other interventions.
- Slow-wave sleep, which is concentrated in the early part of the night, is where memory consolidation and academic learning are locked in; cutting sleep short cuts this process short.
- The single highest-leverage change for most families in this age group is removing all screens from the bedroom and setting a screen curfew sixty to ninety minutes before lights-out.
- Sleep is the overnight repair system for your child's emotional brain; children who sleep well handle frustration, anxiety, and social challenges significantly better than those who are chronically short on rest.

8 The Teenage Brain

Your fifteen-year-old does not come out of his room until noon on Saturday. You have watched him scroll his phone until well past midnight on a Tuesday, drag himself to the breakfast table looking like a ghost, and then somehow function (barely) through a full school day. You have said, more times than you can count, that he needs to go to bed earlier. He has nodded, promised, and done exactly none of it. And now you are wondering whether this is a parenting failure, a character flaw, or something else entirely.

It is something else entirely. And once you understand what that something is, you will probably feel a combination of relief, frustration, and a new kind of low-grade fury directed squarely at whoever decided high school should start at 7:15 in the morning.

Here is what is actually happening inside your teenager's brain. Around the time puberty begins, a biological shift occurs in the circadian clock's timing. The internal body clock, governed by a tiny region of the brain called the suprachiasmatic nucleus, moves its schedule later. Melatonin, the hormone that signals the brain to begin winding down for sleep, starts to rise later in the evening. For most adolescents, this shift is roughly one to two hours later than for a pre-pubescent child or a middle-aged adult. Where a ten-year-old might start feeling genuinely sleepy around 8:30 or 9 at night, a fifteen-year-old's brain will not begin releasing meaningful melatonin until closer to 11 p.m. This is not a choice. It is not laziness. It is not a bad attitude. It is neurobiology doing exactly what it was designed to do at this stage of development.

The problem, of course, is that no one told the school district about the melatonin schedule. That 7:00 a.m. alarm is asking your teenager to wake up at what is, for his brain, the equivalent of 5 in the morning. Every single school day. Week after week. The science on this is not new, not ambiguous, and not small. Chronic sleep restriction in adolescents is one of the most thoroughly documented and most thoroughly ignored public health problems in pediatric medicine. And the parents who are watching this play out in their own kitchens every morning are, in many ways, the closest to the reality of it.

This chapter is for you. Not just to explain the biology, though that matters. But to help you see your teenager differently, to give you language for conversations that tend to go sideways, and to examine what you can and cannot control honestly. Because some of this lives at the policy level, and some of it lives right there in your own home. Let's start with what is happening in the brain.

Chapter Summary

The adolescent years bring a genuine biological shift in sleep timing. At puberty, the body's internal clock delays by one to two hours, pushing the natural sleep window later into the night. This is not a phase teenagers can power through with enough willpower. It is a programmed neurological change. At the same time, adolescents still need eight to ten hours of sleep per night to function well, and the vast majority of American teenagers are getting six to seven hours on school nights. The arithmetic is painful.

This chapter covers the science behind the adolescent phase delay, explains why the "lazy teenager" narrative is biologically unfair, and walks through everything parents actually have some influence over: light exposure, phones in bedrooms, caffeine, the social jet lag trap, and the parent-teen dynamic around sleep rules. It also addresses the mental health connection directly, because the relationship between sleep debt and teen depression and anxiety is real, significant, and worth talking about plainly. School start time policy is part of the picture too, because understanding the landscape helps parents decide whether and how to advocate for change.

What This Chapter Will Cover

- The adolescent phase delay is biological, not behavioral. Melatonin onset shifts one to two hours later during puberty, meaning teenagers are not able to fall asleep at 9 p.m. even if they wanted to. This chapter explains the mechanism clearly so you have something to work with in conversations with your teen, rather than just telling them to try harder.

- Teen sleep need is eight to ten hours per night. The American Academy of Sleep Medicine, the American Academy of Pediatrics, and the Centers for Disease Control all agree on this range. Most American high schoolers get six to seven hours on school nights, and that gap drives much of what parents and teachers observe as moodiness, poor focus, and emotional volatility.

- Early school start times are the structural driver. High schools starting before 8:30 a.m. systematically push teens into chronic sleep restriction. Multiple research programs have tracked what happens when start times are moved later, and the results are consistent: grades improve, mental health indicators improve, and car accident rates among teen drivers drop.

- Weekend recovery sleep and social jet lag create their own problems. Teens who stay up until 2 a.m. Saturday, sleep until noon. Sunday, shift their clocks further, making Monday morning even harder. The chapter offers a practical middle ground.

- Phones in bedrooms are the most controllable family-level factor. The relationship between late-night device use and shortened teen sleep is well-established. This chapter is honest about how hard the conversation is, and why it is worth having firmly.

- Sleep and teen mental health are deeply connected. Sleep debt is not just making teenagers tired; it is amplifying depression, anxiety, and emotional dysregulation in ways that are measurable and serious. Understanding this link changes the frame from a convenience issue to a health issue.

• Caffeine use in adolescents is often underestimated. Many teens consume caffeine in amounts that rival those of adults, and the chapter covers what that means for sleep timing and quality.

Diagram 8.1: The Adolescent Phase Delay

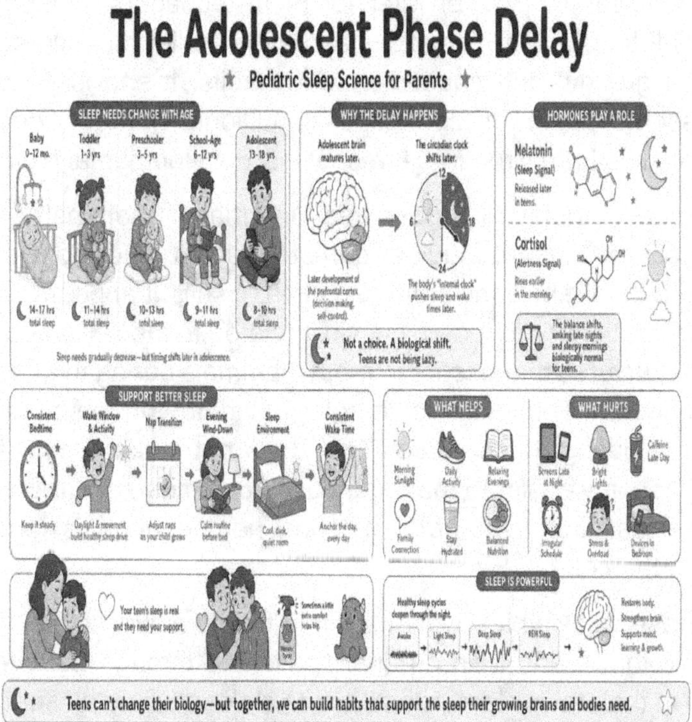

Why the Clock Shifts at Puberty

The circadian clock is not a fixed, unchanging structure. It responds to hormones, light, temperature, and age. During childhood, it tends to run slightly earlier than adult timing, which is why young children often wake up before 6 a.m. and fall asleep easily by 7:30 or 8 at night. Then puberty arrives, and the clock begins a gradual, reliable shift in the other direction.

The mechanism involves multiple hormonal signals, but one of the clearest players is the interaction between puberty hormones and the pathways that regulate melatonin synthesis. The pineal gland, which releases melatonin, becomes responsive to a slightly different light-dark rhythm as puberty progresses. The net effect is that dim-light melatonin onset, the point in the evening when melatonin concentrations begin to rise in a dark room, shifts to approximately 11 p.m. or later for most teenagers, compared to roughly 9 to 9:30 p.m. for adults and even earlier for children. This is not a gradual drift caused by bad habits. It shows up among adolescents living in controlled conditions without access to screens. It is hardwired into the species' developmental biology.

This phase delay is observed in many mammals during adolescence. Rodent studies, primate studies, and cross-cultural human data all show the same direction of shift. It appears to serve some evolutionary purpose, possibly related to the separation from family structures that occurs in most mammals during the transition to adulthood. The brain is literally trying to run on a different schedule during these years. That does not make 7 a.m. school start times any easier to manage, but it does mean that framing the problem as your teenager not trying hard enough is genuinely missing the point.

The phase delay typically peaks in mid-to-late adolescence, around ages 16 to 19, and then gradually reverses through the twenties. Adults in their thirties are often back to a sleep timing closer to their pre-pubescent selves, which is one reason college students look so much worse at morning classes than high schoolers do. By age forty, many people find they are naturally waking up earlier than they ever expected to. Your teenager will almost certainly not believe you when you tell them this, but it is true.

Diagram 8.2: Teen Melatonin Curve vs Adult

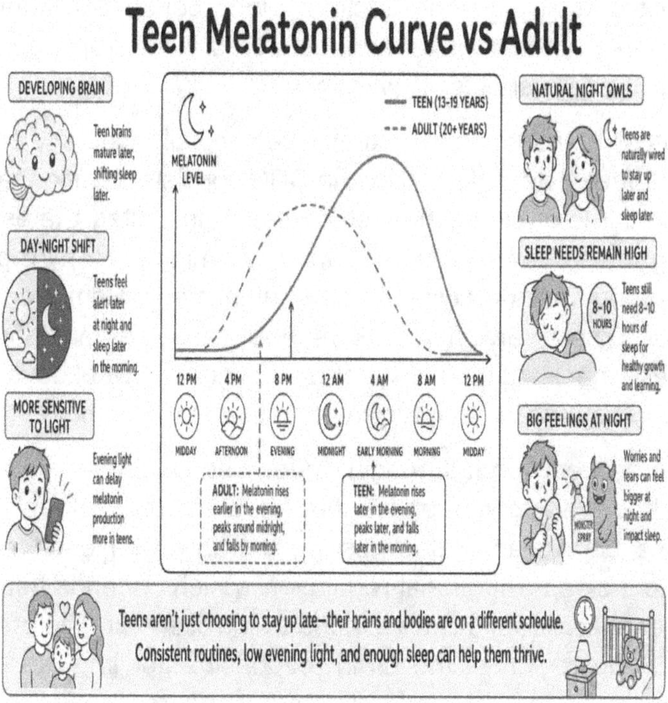

Eight to Ten Hours Is Not a Suggestion

The consensus is unusually tight on this one. When the American Academy of Pediatrics, the American Academy of Sleep Medicine, and the Centers for Disease Control all land on the same number, that is worth paying attention to. For teenagers ages 13 through 18, the recommended sleep duration is eight to ten hours per night. This is not a ceiling; it is a range with a floor. Getting eight hours is acceptable. Getting seven hours consistently is not.

The reason teenagers need so much sleep is that their brains are in an extraordinary period of development. Adolescence is one of the two biggest windows of synaptic pruning and neural remodeling in human development, the first being the first two years of life. During sleep, particularly during slow-wave deep sleep, the brain consolidates learning, clears metabolic waste, and literally reorganizes itself for adult function. Sleep deprivation during this window does not just make teenagers tired. It interrupts a process that is, in some respects, irreversible.

What most teenagers are actually getting falls well short of this target. Survey data collected by the CDC consistently find that more than two-thirds of high school students in the United States report sleeping fewer than eight hours on school nights. Many report six to seven hours as their typical school-night sleep. Some report five or fewer. These are not outliers; they are the majority. Your teenager who is getting six hours a night on weekdays is not unusual. They are, in fact, typical, and that is the problem.

The gap between what teenagers need and what they are actually getting has real consequences that show up in observable ways. Cognitive performance declines measurably with each night of insufficient sleep. Emotional regulation, which depends heavily on a well-rested prefrontal cortex, takes a significant hit. The prefrontal cortex is the region of the brain responsible for impulse control, long-term thinking, and moderating emotional responses. It is also the last part of the brain to fully mature, not completing development until the mid-twenties. Sleep deprivation hits the prefrontal cortex particularly hard, which means a sleep-deprived teenager is functionally more impulsive, more emotionally reactive, and less able to plan than a rested one. This is biology, not attitude.

Diagram 8.3: Teen Sleep Need vs School Start

The 7 a.m. Bell and What It Actually Does

Consider what a 7:15 a.m. school start time asks of a teenager with a natural sleep onset around midnight. If they fall asleep at midnight and need to be awake at 6:30 a.m. to get ready, they are sleeping 6.5 hours. If they fall asleep at 11:30, they get 7 hours of sleep. Neither of these meets the eight-hour floor. And this is not occasional. This is five days a week, 36 weeks a year, across six or seven years of secondary education.

The accumulated sleep debt from this kind of chronic restriction does not just sit quietly. It builds. Cognitive deficits compound. Emotional volatility increases. Immune function is blunted. And the social and academic consequences of all of this show up in grades, in disciplinary records, in athletic performance, and in mental health referrals. The story of early school start times is, in many respects, a story of what happens when an institution is designed for administrative convenience rather than biological reality.

The research on what happens when schools move their start times later is now substantial enough to feel conclusive. Studies tracking districts that shifted start times to 8:30 a.m. or later have found consistent improvements: students slept more, not just by moving their bedtime later, but because the later wake time allowed them to accumulate more of the sleep their bodies were already trying to get. Academic performance improved. Rates of depression and anxiety in student populations declined. Teen car accident rates, which are higher in sleep-deprived drivers, dropped measurably in districts that made the change. These are not small effect sizes. They are large enough that the research case for later start times is essentially settled among sleep scientists.

Diagram 8.4: The AASM Start-Time Recommendation

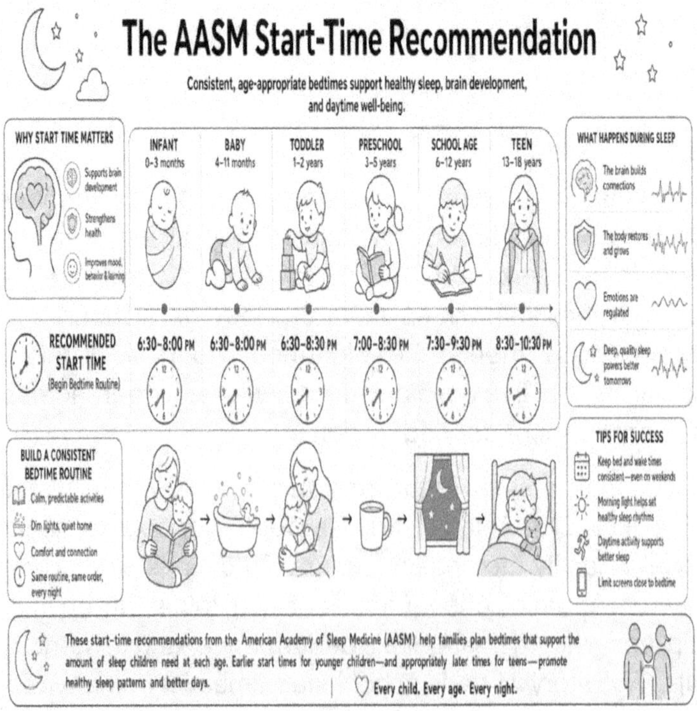

What the AAP and AASM Have Said

The American Academy of Pediatrics issued a policy statement recommending that middle and high schools start no earlier than 8:30 a.m. This was published in 2014 and has been reaffirmed since. The American Academy of Sleep Medicine issued a similar position statement. Both organizations emphasize that the evidence base is strong, the health stakes are significant, and the barriers to implementation are logistical and political rather than scientific.

The barriers are real. Later school start times create scheduling complications for parents who work early, for bus routes, for after-school sports, and for the cascading ripple effects of adjusting one part of a large institution. Some districts have made the change and navigated these challenges. Others have not tried. The point is not that it is easy. The point is that the pediatric and sleep medicine communities have spoken with unusual clarity, and the response at the institutional level has been slow.

For parents, knowing where the official positions land is useful for two reasons. First, it validates what you may already be observing at home. Your teenager is not uniquely struggling. The system is structurally misaligned with adolescent biology, and major medical organizations have formally recognized that misalignment. Second, if you are inclined to advocate for change in your school dictrict, you have authoritative backing. The AAP and AASM positions are public and accessible, and they carry real weight in school board conversations.

What you cannot do, at least not quickly, is change your own school's start time. What you can do is manage everything within your home, and that is where the rest of this chapter focuses.

Light as a Tool, Not Just a Problem

Light is the most powerful signal the circadian clock receives. The suprachiasmatic nucleus tracks the light-dark cycle and uses it to calibrate the timing of melatonin release and the body's overall circadian rhythm. For teenagers, light serves as both a problem and a tool, depending on when it arrives.

In the evening, bright light, and particularly the short-wavelength blue light emitted by phone screens, tablets, and LED lighting, suppresses melatonin release. When your teenager is lying in bed looking at a screen at 10:30 p.m., that screen is actively telling their brain that it is the middle of the afternoon. Melatonin stays suppressed. The already-delayed sleep onset is pushed even further. This is the biological mechanism behind the phone-sleep connection, and it is worth explaining to teenagers directly rather than just issuing rules. When people understand why something matters, they are somewhat more likely to cooperate with it.

Morning light works in the opposite direction. Bright light in the morning, particularly natural sunlight, sends a strong phase-advancing signal to the circadian clock. For a teenager whose clock is running late, getting outside for even 15 to 20 minutes within an hour of waking can help nudge the clock forward slightly over days and weeks. This is not a cure for a delayed phase, but it is a meaningful intervention. Opening the blinds the moment they get up, going outside for a few minutes before school, or even eating breakfast near a window all count. The circadian system is responsive to these signals, and consistency matters more than intensity.

Diagram 8.5: Evening Light vs Morning Light

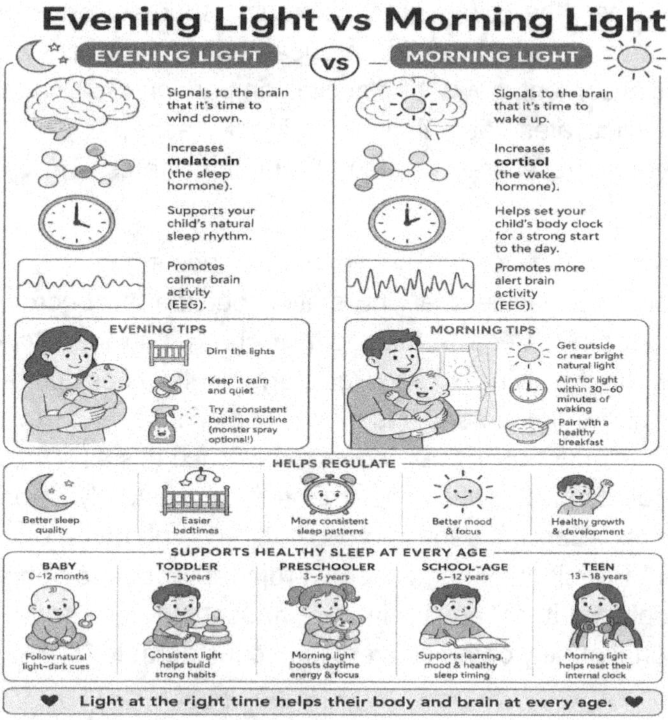

The Phone in the Bedroom

Let's be honest about this one. The phone in the bedroom is probably the single most effective lever most families have over adolescent sleep quality, and it is also one of the hardest conversations to have. Teenagers experience their phones as extensions of their social world, and asking them to disconnect at night feels to many like missing out on something important. That feeling is not irrational. Their social lives genuinely do happen on their phones. The question is whether the cost of those late-night social connections is worth the sleep they take away.

The research on this is not subtle. Studies tracking phone use and sleep duration in teenagers consistently find that later phone use correlates with later sleep onset, shorter total sleep time, and lower sleep quality. The effect appears to be driven by two separate mechanisms: the blue-light melatonin suppression discussed above, and the psychological arousal that comes from engaging with social content, news, and social comparison. Scrolling through a feed activates attention, emotional reactivity, and social evaluation. These are not sleep-promoting states. They are the opposite.

The most effective family approach, according to available evidence, is a device charging station outside the bedroom. The phone does not sleep in the room. It charges in the kitchen, the hallway, or the living room. Resistance will almost certainly be significant the first time this is introduced. Teens who have grown up with their phones in their rooms experience this as a meaningful loss. But families who implement it consistently report that teenagers adapt, sleep improves, and arguments eventually diminish. Framing it as a household rule rather than a punishment, and applying it to adults in the household too, reduces the adversarial quality of the conversation considerably.

Diagram 8.6: The Phone-Bed Pattern

Caffeine and the Teen Brain

Caffeine is the world's most widely used psychoactive drug, and teenagers are using it in quantities that most parents significantly underestimate. The problem is not coffee; it is that the caffeine landscape has changed dramatically in the last decade. Energy drinks now routinely contain 150 to 300 milligrams of caffeine per can, compared to about 95 milligrams in a typical 8-ounce cup of coffee. Large specialty coffee drinks from chains popular with teenagers can contain 250 to 400 milligrams of caffeine. Some teens consume 400 to 600 milligrams of caffeine per day, which is at or above the level considered high, even for adults.

Caffeine works by blocking adenosine receptors in the brain. Adenosine is the chemical that accumulates throughout the day and creates what researchers call sleep pressure, the progressive feeling of sleepiness that builds as waking hours accumulate. Caffeine does not eliminate that sleepiness; it masks it. The adenosine is still there, still building, and when the caffeine clears the body, it returns with a rush. This is the caffeine crash.

The half-life of caffeine in adults is approximately five to six hours. In teenagers, the half-life is similar, meaning that a 300-milligram energy drink consumed at 4 p.m. still has roughly 150 milligrams of active caffeine in the body at 10 p.m. For a teenager already trying to fall asleep on a delayed schedule, that residual caffeine is a genuine obstacle. If they consumed caffeine at 6 p.m. or later, the barrier is even higher.

The conversation about caffeine with teenagers is often more productive than the conversation about phones, partly because the biological logic is easier to grasp. Most teenagers understand at least roughly how caffeine works. Walking through the half-life math together, the idea that the energy drink they had after school is still fighting their sleep at midnight lands for many teens in a way that abstract "you need more sleep" conversations do not.

Diagram 8.7: Caffeine and the Teen Brain

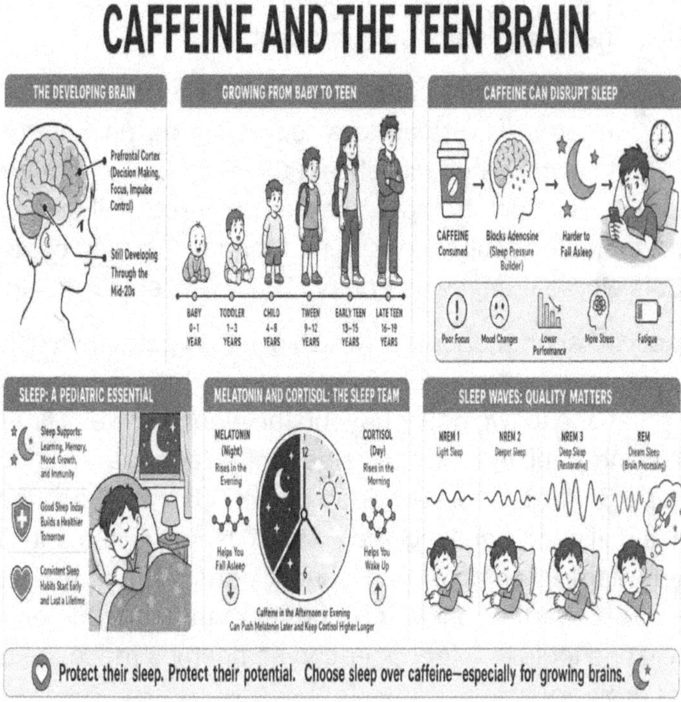

Social Jet Lag and the Weekend Spiral

Social jet lag is a term used in sleep research to describe the mismatch between a person's biological clock and the clock they are forced to live by during the work or school week. For teenagers, it plays out in a pattern that is probably familiar to every parent of a teen: they stay up late on Friday and Saturday nights, sleep until noon or later on weekend mornings, and then drag into Monday morning even more destroyed than usual.

The logic behind this pattern is completely reasonable from the teenager's perspective. They are, finally, sleeping in a way that matches their biology. Friday and Saturday nights, there is no 6:30 alarm looming. They fall asleep when they are actually tired, sleep as long as their bodies want to, and, for the first time all week, something resembling adequate sleep occurs. Of course they want to do this. Their bodies are begging for it.

The problem is that this pattern functions like flying across time zones. A teenager who goes to sleep at 1 a.m. on Saturday and wakes up at noon has effectively set their clock two to three hours further back. By Sunday night, they are trying to fall asleep at 11 p.m., but their clock still reads 2 a.m. Monday morning comes, and the debt is worse than it was Friday. This is the social jet lag spiral.

The practical recommendation, which exists in genuine tension with the desire to let teenagers recover, is to cap weekend sleep extension at one to two hours beyond the weekday wake time rather than allowing a full open-ended sleep-in. A teenager who normally wakes at 6:30 for school can sleep until 8:00 or 8:30 on weekends without significantly shifting their clock. Noon is too far. This is genuinely hard to enforce, and perfect adherence is probably not realistic. But a partial constraint is better than none, and explaining the time-zone logic to a teenager can sometimes make the request feel less arbitrary.

Sleep Debt and Mental Health in Teens

This section needs to be said plainly, because the stakes are real. Sleep and mental health are not just loosely associated in adolescents. They are deeply, bidirectionally connected, and the relationship between chronic sleep restriction and the mental health difficulties that so many families are navigating with their teenagers is substantial.

Teenagers who are chronically sleep-deprived show elevated rates of depression. They show elevated rates of anxiety. They show more suicidal ideation. These are not small correlations. Studies tracking adolescent populations over time find that sleep duration is one of the stronger predictors of mental health outcomes in this age group, comparable in effect size to many other risk factors that receive considerably more attention. A brain that is not sleeping is under metabolic and emotional stress.

The mechanism is not mysterious. The prefrontal cortex, which regulates emotional response and moderates the activity of the amygdala (the brain's threat-detection center), is profoundly sensitive to sleep deprivation. A tired prefrontal cortex cannot do its regulatory job. The amygdala is less inhibited. Threat signals are louder, emotional reactions are bigger, and the ability to contextualize and manage distress is reduced. Sleep-deprived teenagers are not being dramatic. Their emotional systems are running at a structural deficit.

This matters for parents because it reframes what sleep deprivation looks like. The moodiness, the irritability, the tearfulness, the apparent hopelessness that parents sometimes observe in their sleep-restricted teenagers is not just a personality problem. It is, in significant part, a neurological consequence of insufficient sleep. That does not mean sleep is the only factor, or that a teenager struggling with depression will be cured by better sleep hygiene. But it does mean that sleep is a non-negotiable foundation for mental health support in this age group. Getting the sleep right is not separate from addressing mental health; it is part of it.

Diagram 8.8: Teen Sleep and Mental Health

If your teenager is currently seeing a therapist, psychiatrist, or counselor for mood, anxiety, or other mental health concerns, their sleep is worth raising explicitly in those conversations. Sleep assessment is now a routine part of evidence-based adolescent mental health treatment, and many clinicians will prioritize it. If it is not coming up, bring it up yourself.

It is also worth acknowledging that the relationship runs in both directions. Teens with depression and anxiety frequently have sleep difficulties, not just because their mood disrupts sleep but because both conditions are associated with the same neurobiological vulnerabilities. Treating the mood without addressing the sleep, or addressing the sleep without treating the mood, is often incomplete. The two are intertwined, and progress on one front frequently supports progress on the other.

Navigating This with Your Teenager

Here is the honest reality of parenting an adolescent around sleep: you cannot make them sleep. You can create conditions, set policies, and have conversations. But a sixteen-year-old who decides not to comply will not comply, and the parental tools that worked at age six do not work the same way at age sixteen. The goal shifts from control to influence, which requires a somewhat different approach.

Start with the science, not the rules. Many teenagers respond better to honest, direct information about their own biology than they do to being told what to do. Explaining the phase delay, melatonin timing, caffeine half-life, and mental health data can open a different kind of conversation than "you need to go to bed earlier." Teenagers want to understand their own experience. The information in this chapter is not above their comprehension; it is exactly the kind of thing many of them find genuinely interesting.

Then negotiate the practical pieces. Where will the phone charge at night? What is the caffeine cutoff time? What is the weekend wake time limit? Involving your teenager in setting these parameters is not the same as letting them opt out. It means they have some ownership over the rules, which increases the likelihood that they will follow them and also models the kind of self-regulation you are hoping they will eventually internalize.

Diagram 8.9: The Parent-Teen Sleep Negotiation

Recognize what you can control and what you cannot. You can control whether a charging station is available in the kitchen. You cannot control whether your teenager lies awake staring at the ceiling at 11:30 p.m. You can set a household caffeine rule. You cannot supervise every drink they have at school. You can advocate for later school start times. You cannot change the schedule by yourself this year. Working within the real boundaries of influence, rather than fighting battles you will lose, is the more sustainable strategy.

Also recognize that you are allowed to be firm about what genuinely matters. The phone-out-of-bedroom rule is not a suggestion you can offer and walk away from. It is a health rule, backed by evidence, that is worth holding consistently. The same goes for caffeine cutoffs and a reasonable weekend sleep ceiling. There is a difference between respecting a teenager's autonomy and ceding ground on things that directly affect their health and development. You can do both: give them agency in how the rules are set while still holding the line on the rules themselves.

What You Can Do at the School Level

Even if your school district has not moved to later start times, knowing what has been tried and what works gives you something to take to a school board meeting, a parent-teacher organization, or a conversation with your school's principal.

The advocacy argument is stronger when it is grounded in the published positions of major pediatric organizations, which favor start times of 8:30 a.m. or later for middle and high schools. It is also stronger when paired with local data. If your district or school tracks data such as attendance rates, disciplinary incidents, or average GPA by class period, these data can be striking when compared with research findings on sleep-deprived teens. Some districts have also found value in surveying students and parents directly about sleep habits and morning functioning. The results of those surveys, when presented alongside national research, can be compelling evidence in policy conversations.

You may also find allies you did not expect. Many teachers are acutely aware that first-period classes are different from second- and third-period ones. Many school nurses see the physical manifestations of sleep deprivation every day. Coaches notice differences in athletic performance and injury rates that track with sleep schedules. Counselors see the mental health consequences playing out in their offices. Building a coalition around the evidence, rather than approaching it as a single-parent complaint, tends to move things forward in institutional settings. When the PE teacher, the school nurse, the counseling team, and a group of informed parents are all pointing at the same problem, the conversation changes character.

There are also practical models to point to. Several districts across the United States have transitioned to start times later and have carefully documented the outcomes. Their experiences, including the logistical challenges and how they were resolved, are available through organizations like Start School Later, a nonprofit that has been tracking and supporting this work for years. You do not need to reinvent the wheel. The roadmap exists.

None of this will change your teenager's alarm time tomorrow morning. But it is worth understanding that this is not a personal problem unique to your family. It is a structural mismatch between an institution and the biology of the children it serves. Knowing that, and knowing that others are actively working to change it, is at least something to hold onto during the 6:30 a.m. arguments. You are not imagining the problem. You are not alone in it.

The Long View

Your teenager will not be a teenager forever. The phase delay making your mornings so difficult right now is a developmental one. It peaks in mid-to-late adolescence and then gradually reverses. The young adult version of your child, somewhere in their mid-to-late twenties, will probably be the one irritating their own college roommates by wanting to go to sleep by midnight. The biology shifts again, and it shifts back. That does not make right now easier, but it is useful to hold in mind.

In the meantime, your job is to keep the conditions as favorable as you can. Keep the bedroom cool and dark. Keep the phone out of the room. Set a reasonable caffeine boundary and explain why. Help them understand what is happening in their brain rather than framing every sleep conversation as a discipline issue. Know that getting this right has real and lasting consequences for their mental health, their academic performance, and their physical wellbeing, and that the effort you are putting in now, even when it feels like you are fighting the tide, is doing something real and important.

It is also worth noting that habits formed during adolescence, including those around sleep, tend to persist. A teenager who learns to protect their sleep, to understand the role of light, caffeine, and social media in their biology, and to think of sleep as something worth prioritizing is building a foundation they will carry forward. You may not see the payoff immediately. You may not get any credit for it. But the seeds of adult sleep health are planted during these years, and planting them well is worth the effort.

You are not going to solve the structural problem perfectly. No individual family can. But you can reduce the harm considerably, and you can do it in a way that builds a relationship with your teenager rather than fraying it. That combination- honesty about the biology, firmness about the things that genuinely matter, and grace about the things you cannot change- is about the best any parent can do. And honestly, that is quite a lot.

Try This Tonight

For this age range: 13 to 18 years

1. Set up a phone charging station outside your teenager's bedroom tonight. The kitchen counter, a hallway shelf, or a dedicated charging spot in a common area all work. Announce it as a household rule, not a punishment, and plug your own phone in there too.

2. Have a five-minute conversation about melatonin timing. Not a lecture; share one fact: "Your brain doesn't start making the sleep hormone until around 11 p.m., which is why you actually can't fall asleep at 9. That's not a you problem." Most teenagers find this genuinely interesting rather than patronizing.

3. Identify the latest caffeine cutoff you can realistically hold. For most teenagers, 2 p.m. is a reasonable target. If they are currently drinking energy drinks or large coffees at 5 or 6 p.m., moving to 2 p.m. will make a noticeable difference in sleep onset within a week.

4. Open the blinds or go outside with them within 30 minutes of waking up. Morning light is the most direct signal you can give the circadian clock to start running on a slightly earlier schedule. Even a few minutes matter.

5. Set a weekend wake time ceiling. Agree together on a maximum sleep-in time, one to two hours beyond the school wake time rather than a full open-ended morning. Frame it as clock maintenance, not punishment.

6. If your teenager is struggling with mood, anxiety, or emotional regulation, raise their sleep duration explicitly with their healthcare provider or therapist. Sleep is a clinical variable in adolescent mental health treatment, and it is worth tracking and discussing directly.

7. Take one look at the AAP and AASM positions on school start times. If your school starts before 8:30 a.m. and there is any parent advocacy organization at your school or district, you now have the professional backing to raise it as a health issue, not just a preference.

Parent Takeaways

- The teen phase delay is real neurobiology: melatonin onset shifts one to two hours later at puberty, making early sleep onset biologically impossible, not just a matter of effort.

- Teenagers need eight to ten hours of sleep per night; most are getting six to seven on school nights, and that gap has measurable effects on their mental health, emotional regulation, and cognitive performance.

- The AAP and AASM both formally recommend school start times no earlier than 8:30 a.m. for middle and high schools; the evidence base is substantial, and the positions are worth knowing as a parent.

- Phones out of the bedroom is the single most controllable and most effective family-level intervention for teen sleep, and applying the rule to the whole household makes it land differently.
- The link between teen sleep deprivation and depression, anxiety, and emotional dysregulation is real and bidirectional; sleep is not separate from mental health in this age group; it is foundational to it.

9 Routines That Hold

It is 7:14 on a Tuesday evening, and you are already behind.

The one-year-old is melting down in the high chair because dinner ran twenty minutes long. The four-year-old is standing in the hallway in her pajamas, holding her stuffed rabbit and asking, for the third time, whether you are going to read two books or three. The eight-year-old just got home from soccer practice, still in muddy cleats, asking what's for dinner, completely unaware that dinner happened forty-five minutes ago and you have absolutely nothing left in the tank to make another one.

Your partner is at work. They left at four. They won't be home until ten.

You look around at this beautiful, exhausting situation and realize something nobody told you when you were reading sleep books during those quiet evenings of your first pregnancy: the books were mostly written for one child. One age. One bedtime. One neat window of wind-down, a bath, a story, a song, and lights out.

What you have is not that. What you have is a household. A system with multiple moving parts, competing developmental needs, staggered bedtimes that have actually to stack on top of each other, and a parent energy budget that is, to be honest, overdrawn before 5 p.m. most days.

This chapter is for you. Not the idealized version of your family in a controlled sleep study, but your actual family, in your actual house, on a Tuesday.

If you have two children, you already know that the sleep strategies that worked beautifully for your first child had to be rebuilt from the ground up for your second. If you have three or more, you have probably stopped reading sleep books entirely because none of them speak to your reality. If you are a single parent, you have been mentally editing every "one parent handles bedtime while the other one..." sentence since the beginning of this book, and you deserve a chapter that sees you clearly.

And if you are in a two-home co-parenting situation, or you or your partner works a rotating shift, or you are a grandparent raising grandchildren, this chapter will try to speak to you too.

Bedtime in a real household is not a single event. It is a choreography. And once you start thinking of it that way, things get a little easier to manage.

Chapter Summary

This chapter is about building bedtime systems that actually hold across the full complexity of your household. It takes the science from earlier chapters and puts it into real-world architecture, covering multi-child homes, single-parent households, shift-work schedules, and co-parenting across two homes.

The core idea is this: you are not just trying to get one child to sleep. You are trying to move multiple humans with different sleep needs through a shared wind-down window without anyone unraveling. That requires a different kind of planning, one that thinks at the household level rather than the individual child level. This chapter gives you the frameworks, the flexibility, and the honest acknowledgment that some nights none of it goes according to plan.

What This Chapter Will Cover

- Why generic schedules fail real families. Most sleep advice assumes a single child, two present parents, and a predictable evening. Real households do not look like that. This section explains why one-size-fits-all bedtime advice breaks down and what to build instead.
- Staggered versus synced bedtimes. Deciding whether your children go to bed at the same time or in sequence is one of the biggest structural decisions in a multi-child household. The

answer depends on age gaps, parent bandwidth, and your home's layout. This section walks through how to make that call.

• The three-or-more-kid problem. When you have three or more children, tag-team handoffs and zoning strategies become essential. This section outlines how to divide the choreography between parents or sequence it solo as needed.

• The single-parent reality. When there is no one to hand off to, the system has to be extremely well-designed. This section addresses decision fatigue, how to build structural support into your routine, and why needing outside help is not a personal failing.

• Shift-work and rotating schedules. When one parent works nights, or alternates between day and night shifts, the household sleep architecture has to flex. This section covers how to protect children's sleep consistency even when the adults' schedules are unpredictable.

• Co-parenting across two homes. Children can sleep well in two households, but it requires intentional coordination between parents. This section offers a framework for keeping enough consistency to protect sleep without requiring identical routines in both homes.

• Travel, holidays, and recovery. Disruption is inevitable. This section covers how to build elasticity into your routine so that a week at grandma's house or a cross-country flight does not undo months of progress, and how to use the 3-night reset when you get home.

Why Generic Schedules Fail Real Families

If you search "toddler bedtime routine" right now, you will find hundreds of articles describing the same sequence: bath, pajamas, brush teeth, one story, one song, lights out at 7:30. It is good advice for one child in stable circumstances. It is almost useless as a complete system for a household with real complexity.

Generic sleep schedules fail real families for a simple reason: they are designed for the average, and no actual family is the average. They assume two parents are both home and available each evening. They assume all children in the household are the same age, or close enough that one bedtime window serves everyone. They assume the evening is roughly predictable, that dinner ends on time, that nobody has an evening activity, and that the only variable is what time you start the bath.

Diagram 9.1: Staggered vs Synced Bedtimes

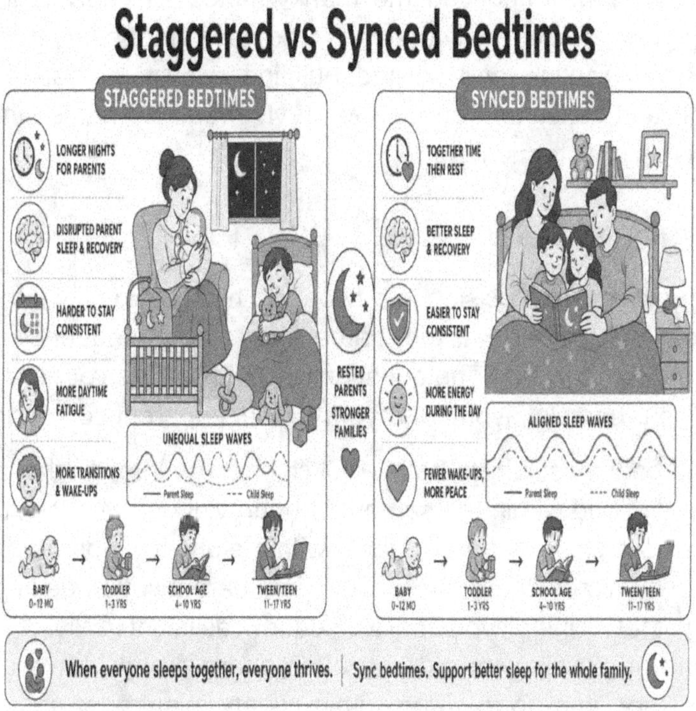

The reality is that most families with more than one child are managing bedtimes across a significant developmental range: a newborn and a four-year-old need completely different things at the end of the day. A toddler and a seven-year-old have circadian systems that are genuinely offset by one to two hours. If you try to put them both to bed at the same time, one will always be overtired, and the other undertired, and neither will fall asleep cleanly.

What families need is not a single bedtime but a bedtime architecture. That architecture accounts for developmental stages, parent availability, home layout, and the kind of disruptions that actually happen in your specific household. It is less like a recipe and more like a floor plan. It tells you who goes where, in what order, for how long, and what happens when someone goes off-script.

The families who find bedtime manageable, not every night, but most nights, are generally the ones who have stopped trying to find the perfect routine and started building a system that is good enough and consistent enough to survive reality. That is what the rest of this chapter is about.

Staggered vs Synced Bedtimes

The first structural decision in a two-child household is whether the children go to bed at the same time or in sequence. Both approaches can work. The question is which one fits your children's ages, your home's layout, and your available parent coverage.

Synced bedtimes, where both or all children go through the routine and land in bed within the same half-hour window, work best when children are close in age (roughly within eighteen months of each other), when they share a room, or when parent coverage is limited enough that running two separate sequences is not realistic. A four-year-old and a six-year-old can often share a bath, have a short group story, and go to sleep within twenty minutes of each other without either one being significantly overtired or undertired.

Staggered bedtimes work better when there is a significant age gap, typically two years or more, and when the older child's later bedtime can be honored without keeping the younger one up too long. A two-year-old and a seven-year-old have genuinely different sleep pressure systems. The two-year-old may need to be asleep by 7:00

or 7:30. The seven-year-old can comfortably push to 8:30 or 9:00 and will actually fall asleep more easily if given that later window.

Diagram 9.2: The Three-Kid Tag-Team Map

The practical challenge with staggered bedtimes in a two-parent household is that someone has to occupy the older child while the younger one is being settled. This is where having a clear "older child anchor" becomes useful. The older child is given something quiet and independent to do, a puzzle, a chapter book, a favorite podcast, while the younger child is walked through bath and wind-down. Then the older child gets their own condensed version of the routine. It does not have to be elaborate. Fifteen minutes of reading together and a short check-in before lights out can be deeply satisfying for a seven-year-old who otherwise feels like they always have to wait.

In single-parent households, the staggered approach often requires one child to be largely self-managing during the other child's routine. Older children can learn to shower independently, pick out their own pajamas, and start reading while a younger sibling is being put to bed. This is not asking too much of a child who is eight or nine. It is actually a developmentally appropriate level of independence, and most kids that age feel proud of it when it is framed as a sign of being grown-up rather than being deprioritized.

The Three-or-More-Kid Problem

When you have three or more children, the logistics of bedtime become genuinely complex. You are no longer just sequencing two sleep windows. You are managing three or more developmental needs, potentially spanning the range from infant to preteen, with a finite amount of parental attention that does not divide cleanly.

The two approaches that work best for larger families are tag-team and zoning.

Tag-team means that two parents divide the children between them, each taking one or two through the routine simultaneously. Parent A handles the youngest two while Parent B handles the older one. Or Parent A bathes all three while Parent B preps the rooms, then Parent A does the baby's final feed while Parent B reads to the middle two. The key to tag-team is that the handoffs are planned, not negotiated at 7:00 p.m. when everyone is already fraying. A quick five-minute conversation in the afternoon ("You take the baby tonight, I'll do the big two?") prevents enormous amounts of friction.

Diagram 9.3: Single-Parent Bedtime Choreography

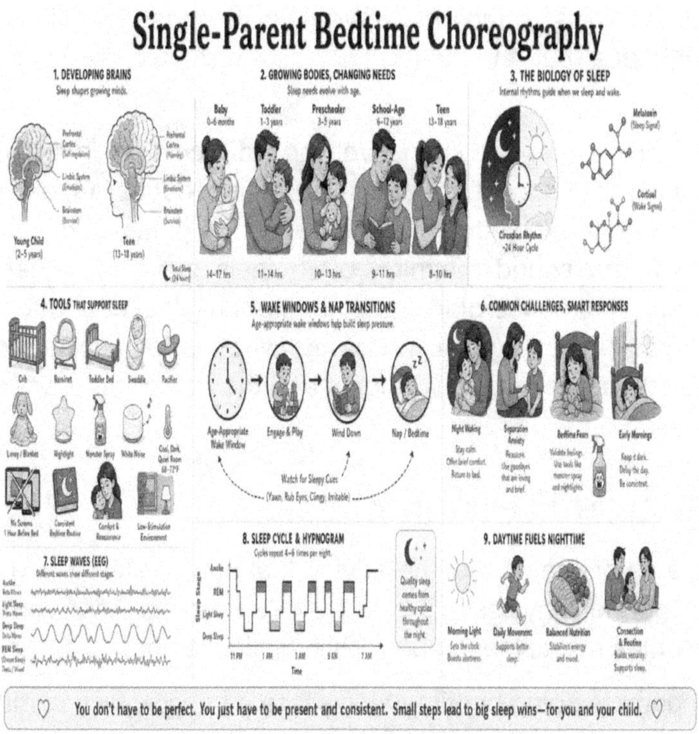

Zoning means dividing the house, not the children. One parent owns the upstairs; the other owns the downstairs. The children assigned to each zone follow their routines in parallel, with minimal cross-contamination. This is particularly useful when the noise of one child's wind-down disrupts another's. The baby's white noise machine in the nursery can drown out the older kids' bath without anyone having to whisper-shout through walls.

In households where only one parent is available on a given evening, the three-kid bedtime routine has to be carefully sequenced to avoid simultaneous meltdowns. The general principle is youngest first when possible, because younger children have less capacity to wait and less ability to self-regulate during a long delay. But this is not an absolute rule. An eight-month-old who has had a late nap may be able to play independently in a safe space for twenty minutes while a three-year-old gets a bath, while a seven-year-old showers on their own. What you are managing is not just the sequence but the timing of each child's peak sleepiness window.

There is also a real benefit to involving older children in the routine of a younger sibling, not as substitute parents, but as helpers. A six-year-old who is allowed to pick out the baby's pajamas or push the button on the sound machine often feels included rather than displaced. That sense of inclusion matters a great deal to children in the middle of a large sibling group, who can otherwise feel lost between the needs of the youngest and the perceived privileges of the oldest.

The Single-Parent Reality

There is a version of bedtime advice that only makes sense if you have a partner physically present every evening. That version of reality writes most sleep books. This section is not.

Single parents doing bedtime alone face a specific kind of difficulty that is not just about being tired, though the tiredness is real and significant. It is about decision fatigue. Every decision about the routine, every deviation, every night when things go off-script, lands on one person. There is no one to look at and ask, "Should we just skip the bath tonight?" There is no one to take over at 9:00 p.m., even though you have been on since 6:00 a.m. The mental load of being the only adult in the house during bedtime is different in kind, not just degree, from sharing it.

Diagram 9.4: Shift-Worker Family Schedule

The most effective thing a single parent can do is build a routine that is extremely predictable and requires very few in-the-moment decisions. When the routine is automated, when every child knows exactly what comes next without being told, the cognitive demand on the parent drops significantly. This means drilling the routine for a few weeks until it becomes almost invisible. The children become the structure. They know that after bath comes pajamas, that after pajamas comes teeth, that after teeth comes one story, that after one story comes lights out. They are not waiting for you to tell them what's next.

Outside support is also structural for single-parent households, not optional or occasional. This does not mean you need a full-time babysitter. It means identifying one or two people who can be present for bedtime one or two nights a week, whether that is a grandparent, a neighbor, an older sibling, or a co-parent in a shared custody arrangement. The goal is not to hand off bedtime entirely, but to build in one or two evenings per week when you are not doing it all alone. Even one covered night changes the sustainability of the whole system.

It is worth saying clearly: needing help is not a personal failing. It is a structural reality. Human children have historically been raised by networks of adults, not by a single person operating in isolation. The nuclear family model is relatively recent and, in many ways, not particularly well suited to the demands of early childhood caregiving. If you are raising children alone and exhausted, that is the expected outcome of an objectively difficult situation, not a reflection of your capabilities.

Shift-Work Families

A family in which one parent works a rotating shift or nights on a regular schedule has to solve a different version of the bedtime problem. The household rhythm cannot be built around both parents being home every evening, because that is not the reality. But children's sleep needs are consistent regardless of the parent schedule. Your toddler's circadian clock does not know that you work nights on Thursdays and days on Fridays.

The most important thing for a shift-work household to protect is the child's anchor, the fixed points in the routine that do not change regardless of which parent is home or what time. The bath happens at the same time. The story happens in the same order. The light goes off at roughly the same clock time every night. The person running the routine changes, but the routine itself stays stable.

Diagram 9.5: Two-Home Co-Parenting Routine

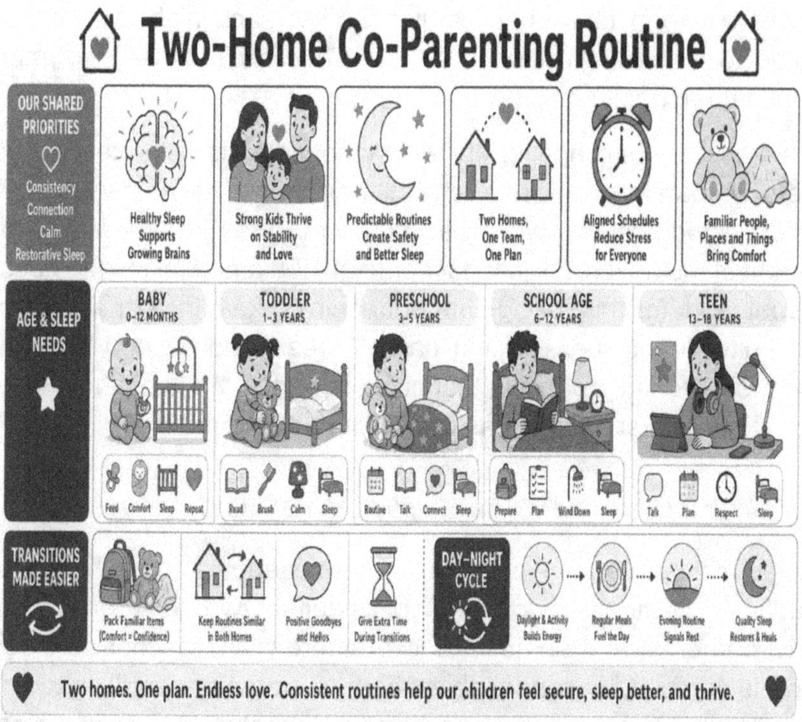

Children are more adaptable than adults in one specific way: they calibrate to place and to routine rather than to person. A four-year-old will go through bedtime smoothly with Dad on Tuesday nights if Dad follows the same sequence that Mom uses on Monday nights. The child is not attached to the parent who does the routine; the child is attached to the routine itself. This is genuinely good news for shift-work families, because it means the handoff can work as long as the departing parent documents the routine clearly enough for the arriving one.

Where shift-work families most often run into trouble is during the re-sync after a stretch of off-pattern days. If a parent works a rotating schedule and has been on nights for four days, the children may have drifted. Weekend re-syncing, pulling bedtime back to its normal window over two or three days rather than trying to snap back immediately, is more effective than a hard correction. Push bedtime fifteen to twenty minutes earlier each night rather than trying to move it an hour at a time. The child's circadian system needs time to follow.

A particular challenge in shift-work households is protecting the sleeping parent's daytime rest. Children do not have an instinct for quiet. A toddler who wakes at 6:00 a.m. does not know that the parent in the next room just got home at 4:00 a.m. Making the house layout work for this, using sound machines, keeping certain rooms off-limits during the sleeping parent's rest window, and involving older children in protecting that quiet, is part of the household sleep architecture just as much as the children's bedtime is.

Co-Parenting Across Two Homes

When children move between two households, they have two bedtime routines, two sets of adults, two different home environments, and sometimes two different approaches to sleep. This is completely manageable if both co-parents are willing to hold enough consistency between homes to protect the child's sleep system. It does not require identical routines. It requires enough overlap that the child's circadian anchor is not disrupted.

The minimum viable consistency for co-parenting sleep is bedtime within thirty to forty-five minutes of the same clock time in both homes, a similar pre-sleep wind-down (at least twenty minutes of calm before lights out), and the same basic sequence of hygiene and settling steps—everything else, which stuffed animals, which books, which parent sits where, can differ. Children are quite capable of holding "we do it this way at Dad's house and that way at Mom's house" as a normal part of their world.

Diagram 9.6: Travel and Time-Zone Recovery

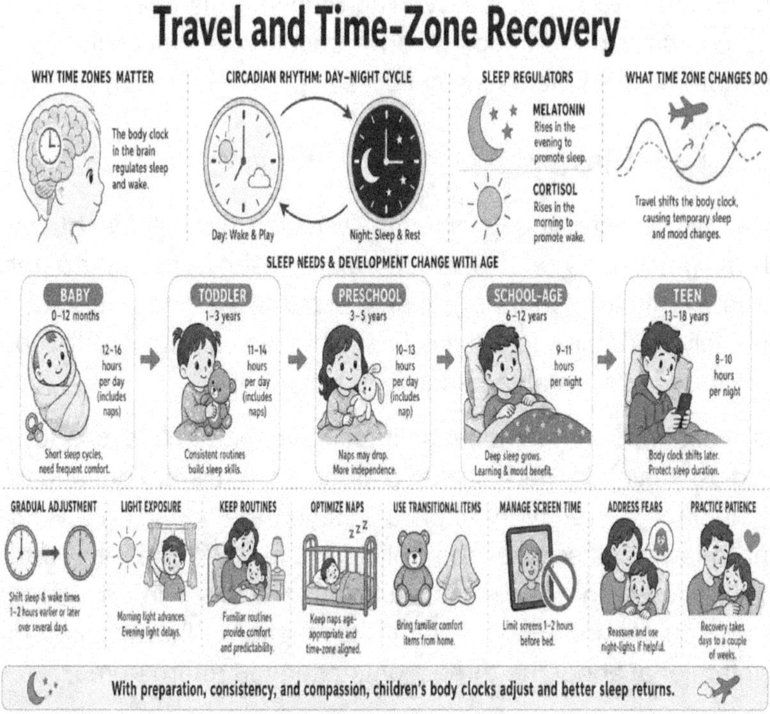

Where co-parenting sleep falls apart is when one household uses the bedtime window as a point of difference, a way to offer more indulgence as a way of compensating for absence or guilt, or, in higher-conflict situations, as a deliberate way to disrupt the child's sleep at the other home. These dynamics are real and are worth naming here. Children whose sleep is disrupted by parental conflict experience consequences for their mood, behavior, and development. If you are in a co-parenting situation where the sleep piece has become a battlefield, the most important thing you can do for your child is find one thing to agree on and hold it, even if it is just the bedtime hour.

For families with a cooperative co-parenting relationship, a shorthand note or a co-parenting app entry on transition days is enormously helpful. "She took a late nap, had a hard afternoon, was up at 5:30 this morning" tells the receiving parent everything they need to know to adjust the evening routine. This does not require deep coordination or ongoing conversation. It is simply sharing the information that protects the child's sleep.

Travel, Holidays, and Grandparent Visits

Every family with young children will face disruptions to travel, holiday gatherings, and extended visits to or from grandparents. These events tend to be joyful and also, from a sleep perspective, genuinely destabilizing. That is fine. Children can handle disruption when they have a stable baseline to return to.

The most important mindset shift for travel and holidays is to stop trying to maintain perfect sleep during the disruption and start planning for the recovery afterward. If you are spending a week at your in-laws' house over the holidays, there will probably be late nights, overtiredness, extra stimulation, and sleep in an unfamiliar place. You cannot fully prevent the disruption. What you can do is be intentional about the three nights after you return home.

Diagram 9.7: Bedsharing Safety Decision Tree

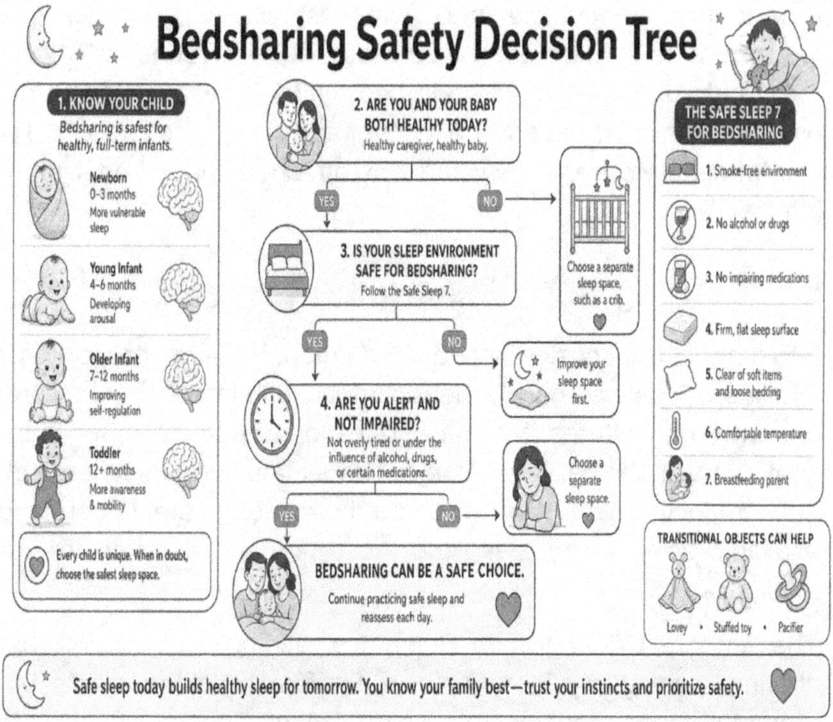

Time-zone travel adds a specific layer of circadian disruption on top of the general stress of unfamiliar sleep environments. For every hour of time zone change, children typically need about 1 day to adjust, similar to adults, though it's often faster in younger children. Flying east is generally harder than flying west, because you are asking the child to fall asleep before their internal clock is ready. Prioritizing morning light exposure and getting your child outside as early as possible in the new time zone helps reset the circadian clock faster than almost anything else.

Grandparent visits bring their own kind of sleep disruption because grandparents often have strong feelings about sleep rules, usually favoring more flexibility. They want to let the baby sleep in their arms. They want to let the toddler stay up so they can spend time together. They were not sleeping by 7:30 when you were little, and you turned out fine. This is a real conversation that most parents have to navigate multiple times. The most effective approach is not to frame it as a rules debate but as a logistics request: "It would really help us if she could be in bed by 7:30, because if she gets overtired it takes three days to recover and we're here for four." That kind of concrete framing, with a why attached, lands better than a blanket rule.

Bedsharing and Cosleeping

Bedsharing and cosleeping are common. A large proportion of families worldwide share a sleep surface with infants and children, and in many cultures it is the unquestioned default rather than a controversial choice. If you are in this situation, this section is not going to lecture you. It will give you the clearest safety information available so you can make the most informed decision for your family.

The American Academy of Pediatrics recommends against bed-sharing, particularly for infants under four months, infants born prematurely, or infants with low birth weight. The specific risks the AAP identifies are related to accidental suffocation, particularly on soft surfaces and in situations where a parent has consumed alcohol, sedating medication, or is extremely sleep-deprived. The recommendation is that infants sleep on their own firm, flat surface, in the same room as the parents but not on the same sleep surface. This is the safest known configuration for the infant sleep period.

Diagram 9.8: The 3-Night Reset

At the same time, the AAP acknowledges that many parents fall asleep while feeding or caring for an infant, and that it is safer to plan for that possibility than to pretend it will not happen. If you know there is a chance you will fall asleep while feeding in bed, the safest thing is to prepare the sleep surface in advance: remove soft bedding and pillows from around the area where the infant will be, make sure the mattress is firm, and ensure there is no gap between the mattress and the headboard or wall where a small body could become wedged.

For toddlers and older children, the safety calculus changes considerably. A healthy two-year-old who cosleeps in a standard adult bed does not face the same suffocation risks as a newborn. The question shifts from acute safety to longer-term considerations about sleep consolidation and family sleep quality. Some families find that cosleeping is genuinely restorative for everyone. Others find that the presence of a child in bed significantly fragments parents' sleep. Neither experience is wrong. If the arrangement is working for your family, including the adults, there is no urgency to change it. If it is not working, later chapters on transitioning older children to independent sleep offer a gradual, respectful approach.

The core message here is: no judgment, no shame, and full access to the safety information you need. You are making decisions for your family in your real circumstances.

Recovery After a Disrupted Week

Every family hits a stretch of disrupted sleep. Illness. Travel. A holiday week that ran too long. A baby who had a growth spurt and stopped sleeping through the night again. A toddler who had a developmental leap that blew up two months of progress in three days. These periods are not failures. They are normal.

The question is how to recover efficiently when the disruption ends.

Diagram 9.9: Routines That Survive Real Life

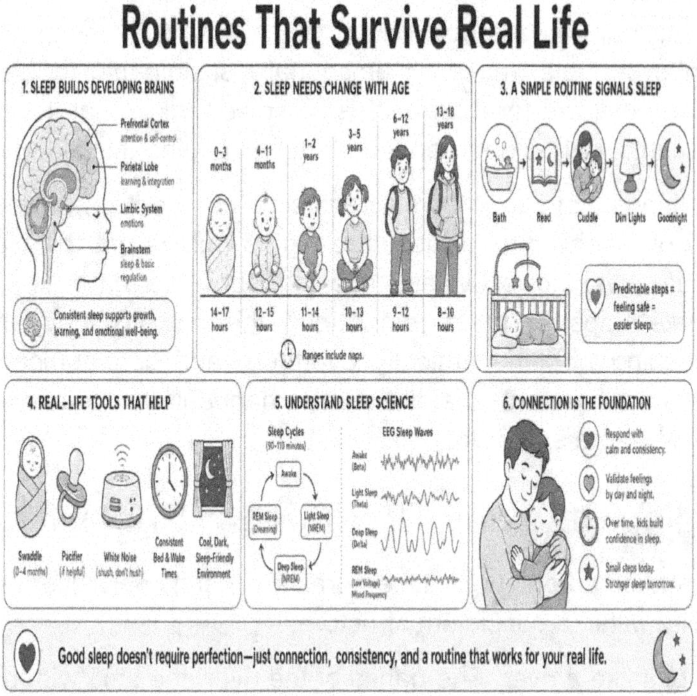

The 3-night reset is a structured approach to re-establishing baseline sleep after disruption. The idea is that for three consecutive nights, you hold the routine with unusual consistency: same bedtime, same sequence, no deviations, gentle but firm responses to any testing or extra requests. You do not introduce new elements during the reset. You do not attempt sleep training or any major change. You are simply restoring the familiar pattern through enough repetition for the child's system to recalibrate.

Night one of the reset is usually the hardest. The child has had a week of exceptions and will push to extend them. Night two is typically easier. Night three, in most cases, looks very close to pre-disruption baseline. This is not magic. It is the simple biology of habit and circadian regularity. The sleep system is not permanently broken after a bad week. It just needs consistent input to find its rhythm again.

What makes the 3-night reset work is its time-bound nature. Knowing that you are committing to three hard evenings, not an indefinite struggle, makes it emotionally sustainable for parents. You can hold firm for three nights when you know that the goal is three nights, not an open-ended stretch of hoping things improve.

A few things that support a successful reset: prioritizing sleep over other obligations during the three nights (this is not the week to schedule evening activities), making sure the child is getting adequate naps or quiet rest during the day so they are not approaching bedtime critically overtired, and communicating to partners or co-parents that the reset is happening so that everyone holds the same standard.

Building Routines That Survive Real Life

What makes a bedtime routine durable is not its perfection. It is its flexibility within a consistent structure.

Think of it like a river. The banks of the river, the time window, the basic sequence, the sleep environment, are consistent. They do not move very much. But within those banks, there is room for water to flow in different ways. Some nights the bath is short. Some nights there are two stories instead of one. Some nights, a child needs an extra five minutes to talk through something that happened at school. The structure holds, even when the water moves differently within it.

The families who find bedtime the most sustainable are generally those who have internalized this distinction. They are not trying to run the same program every night. They are holding the frame while allowing flexibility inside it. The child knows when bedtime is happening. The child knows the sequence. The child knows what comes after lights out. Within that knowing, there is room for the actual texture of a real evening.

This also applies to parents. You will have evenings when you are too tired to do the full routine beautifully. Those nights exist. What matters is that you do it anyway, in whatever abbreviated form is possible. A three-minute abbreviated wind-down is vastly better for your child's sleep system than no wind-down at all. The consistency of the signal, even when it is shorter than usual, trains the brain to associate that sequence with sleep. You do not have to be perfect. You have to be present and consistent enough.

If you have built your bedtime routine well, it will do something remarkable over time: it will start running on its own. The child will remind you of what comes next. The older child will help the younger one with the steps they know by heart. The routine will become part of the household culture rather than something you have to do consciously each evening. That is the goal. Not a perfect system, but one that is alive enough to keep itself going even on the hardest days.

Real families have unexpected work calls. They have sick kids who generously share their illness. They have days where dinner burns and someone cries over something invisible and the whole evening slides sideways. The routine that survives real life is not the one that accounts for none of this. It is the one that bends without breaking, with enough elasticity built in to absorb a hard night and return to baseline by the next.

You are building something that has to work not just on the good nights but on the genuinely terrible ones. That is a high standard. But it is also, when you get there, one of the most stabilizing things in a young child's life. A bedtime routine that holds is not just about sleep. It is about safety and predictability. It is the child knowing, at the end of every day regardless of what that day held, that this sequence of familiar things will happen, that the same voice will read the same book in roughly the same way, and that sleep will come.

That is worth building carefully. That is worth protecting.

Try This Tonight

For this age range: all ages, adapt steps to your children's specific ages and household configuration.

1. Write down the exact sequence of your current bedtime routine for each child, from the moment dinner ends to the moment the light goes off. If you have more than one child, write a separate sequence for each. This is not about judging what you are doing. It is about seeing clearly what the routine actually is versus what you intend it to be.

2. Identify the one moment in the current routine that most reliably goes sideways. The bath that turns into a battle, the "one more story" loop that adds twenty minutes, the older child who always reappears after lights out. Name it specifically. You cannot fix what you have not named.

3. If you have a co-parent, have a five-minute conversation today about whose turn it is for bedtime for the next three nights. Put it on the calendar or in a shared notes app. You are not solving everything tonight. You are just deciding who is on duty.

4. If you are a single parent, identify one night in the next two weeks where you can have another adult present for bedtime, whether that is a family member, a friend, or a paid sitter. One covered night is enough to make the rest of the week feel more manageable.

5. If you have multiple children at different ages, decide tonight whether you are running staggered or synced bedtimes. If you have been ambivalent or inconsistent about this, pick one and try it for a week. A consistent imperfect approach beats an inconsistent attempt at perfection.

6. If your routine has been disrupted recently, tonight is night one of the 3-night reset. Do the routine in full, in sequence, and hold the line gently when tested. Tomorrow will be easier. The night after that will be easier still.

7. End the evening with a brief moment for yourself before you go to sleep, even five minutes. You have been the anchor for your children's sleep system. Your own sleep matters too. The chapters ahead include more on parents' sleep health, but for tonight, allow yourself to close your eyes without immediately running through the list of what still needs to be done.

Parent Takeaways

- A bedtime routine for a multi-child household is a choreography, not a single event. Design it at the household level rather than on a child-by-child basis.
- Stagger bedtimes by thirty to sixty minutes between consecutive age groups once children are past infancy. Trying to sync children with a significant age gap almost always leaves one of them behind.
- Single parents need structural outside support built into the routine, not as an occasional luxury but as a regular part of the system. One or two covered evenings per week changes the long-term sustainability of everything.
- When bedtime has been disrupted by travel, illness, or holidays, use the 3-night reset on return. Three nights of a full, consistent routine restore baseline far faster than gradual drift back does.
- Co-parenting sleep works when bedtime falls within thirty to forty-five minutes of the same clock time in both homes, and the basic wind-down sequence is similar. Identical routines are not required.
- The most durable routines are ones with consistent structure and built-in flexibility. Hold the frame, allow the texture inside it to vary. That is how you build something that survives real life.

What About Screens at Bedtime

It is hard to talk about real household routines without talking about screens. Most children over the age of two are surrounded by screens, and many parents use screens as a wind-down tool either intentionally or out of necessity. This section won't tell you to eliminate all screens before bedtime. It will give you the research and let you make the call.

The core issue with screens in the hour before sleep is the blue light they emit. Blue-wavelength light is highly effective at suppressing melatonin, the hormone that signals to the brain that it is time to sleep. Even relatively short exposure (twenty to thirty minutes) in the hour before bed can delay sleep onset by thirty minutes to an hour in school-age children, and the effect is more pronounced in adolescents whose melatonin onset is already shifted later. This is not a theoretical concern. It is a measured physiological response.

The secondary issue is content-related arousal. High-stimulation content, action sequences, and comedy that elicit strong emotional responses, as well as social media for older kids, activate the stress-response system in ways incompatible with smooth sleep onset. A child who has been watching intense or emotionally engaging content for an hour before bed has a neurologically aroused brain that needs additional time to return to baseline before sleep becomes possible.

That said, context matters. A child watching a slow, familiar, low-stimulation show they have seen many times before is in a meaningfully different situation than one watching something new and exciting. Repetition dampens the novelty response. Some families find that a single familiar episode of a calm show serves as a reliable part of the wind-down routine precisely because it is so predictable. If that is working for your family and your child is falling asleep within a reasonable window, the theoretical concern about blue light may be less important than the practical reality that the routine functions.

The clearest guidance from sleep researchers is to create a buffer of at least 30 minutes, preferably 60, between screen-off and lights-out, and to pair the screen-off moment with a transition to lower-stimulation activity. That transition is easier when it is consistent. "Screen off, then bath, then story" becomes a conditioned sequence that the brain learns to associate with sleep. The screen-off moment stops feeling like a punishment and starts feeling like the beginning of the wind-down.

For older children and teenagers, the conversation about screens is more complicated because their devices are also tools for social interaction, homework, and identity. Taking the phone away at bedtime can feel to a teenager like being cut off from their entire social world. A negotiated agreement, a specific time when the device goes to a charging station outside the bedroom, is more sustainable than a hard confiscation and tends to produce less conflict. The goal is not perfect compliance. It is a workable arrangement in which the teenager has some investment in honoring it.

10 Anxiety, Worries, and the Bedtime Mind

Your nine-year-old has been crying at bedtime for three weeks now. Not big, dramatic sobs, but quiet tears she wipes away fast when you check on her. She says she is not scared of anything specific. She is not asking for a glass of water or a third trip to the bathroom. She is "just worried." You sit on the edge of her bed and ask what about, and she shrugs, looking genuinely bewildered herself, like the worry is too big and too formless to name.

You turn the light back off and pad down the hall, and you hear her shift, sigh, and shift again. An hour later, you check: still awake, staring at the ceiling. You feel the pull of frustration and the sharper pull of worry about her worry. You wonder if something happened at school. You wonder if you missed something. You wonder if this is just a phase or the beginning of something that needs help. You wonder how long you can both keep going on this much lost sleep.

Here is what you need to know right now: what your daughter is experiencing is incredibly common, it has a real neurological explanation, and some tools genuinely work. The fact that bedtime is when her worries finally surface is not a coincidence, nor is it a character flaw in her or a failure in your parenting. It is the predictable outcome of how the brain quiets down after a busy day. Once you understand why it happens, the how-to-help part becomes much clearer.

Bedtime anxiety is one of the most frequent concerns parents bring to pediatricians, sleep specialists, and child therapists. It spans every age from toddlerhood through the teen years, and it wears different faces depending on how old a child is. A two-year-old cries when you leave the room. A five-year-old worries about monsters under the bed or whether the smoke detector will go off. An eight-year-old rehearses tomorrow's spelling test in the dark. A twelve-year-old replays a social awkward moment from lunch over and over. A fifteen-year-old lies awake thinking about college applications, climate change, and a fight with her best friend.

The common thread is this: nighttime is when the brain finally has space to feel things it was too busy to feel during the day. And for children whose nervous systems run a little hot, who are more sensitive, more empathic, or simply going through a hard season, that quiet time becomes a floodgate.

This chapter is for you if you have a child who worries at bedtime in any of the following ways. It will explain the neuroscience behind what is happening in your child's brain, walk you through the most effective strategies that cognitive behavioral therapy offers for children, and help you figure out when worry has crossed from normal into something that needs professional support. Most importantly, it will help you feel less alone in this, because nighttime anxiety in children is one of those parenting experiences that can feel very isolating, very confusing, and very exhausting all at once.

You are doing the right thing by reading this. Let's figure it out together.

Chapter Summary

This chapter addresses the cognitive and emotional roots of children's sleep difficulties, specifically the worries, fears, and bedtime-specific dread that appear in different forms across all age groups. Drawing on age-appropriate adaptations of Cognitive Behavioral Therapy for Insomnia, it provides parents with a clear set of tools: designated worry time earlier in the evening, worry-externalization techniques, breathing- and body-based calming practices, and gentle visualization scripts that children can use independently.

You will learn why anxiety surfaces specifically at bedtime, what developmental fears are normal at each age, and how to tell the difference between everyday worry, situational anxiety during hard seasons, and the kind of persistent anxiety that benefits from professional support. You will also learn about your own role in the equation, because a parent's physiological state during a difficult bedtime is not invisible to a child. Your calm, when you can find it, becomes one of your most powerful tools.

What This Chapter Will Cover

- The neuroscience of bedtime anxiety. When the day's stimulation fades, the brain's threat-detection center activates, and the mind turns inward. Understanding why this happens helps parents respond without panic or frustration and helps older children better understand their own minds.

- Developmental fears by age. Separation anxiety, fear of the dark, monster fears, existential worry, social anxiety, and performance stress are all developmentally sequenced. Knowing what is normal at each age prevents over-pathologizing and helps you meet your child exactly where they are.

- CBT-I principles adapted for children. Stimulus control, gentle sleep restriction, and cognitive restructuring can all be modified for kids. These techniques work, and they are not harsh when applied thoughtfully.

- The worry-time technique. Moving worry conversations to earlier in the evening measurably changes bedtime chemistry. This section walks through exactly how to run a worry-time slot that actually works.

- Breathing, body scans, and progressive muscle relaxation. These are not just placeholder activities. When practiced consistently, they activate the parasympathetic nervous system in real time. This section provides practical scripts for different ages.

- Visualization scripts and worry tools. The safe-place visualization and the worry-jar method are two concrete scripts you can use with your child tonight. Both are age-adaptable and work for children who are too young for purely verbal approaches.

• When to involve a professional. Persistent anxiety, somatic symptoms, school avoidance, and signs of depression all warrant a different level of support. This section provides clear criteria, so you know when to make that call.

Why Bedtime Activates the Anxious Brain

During the day, your child's brain is flooded with input. School, friends, homework, screen time, activity, chatter, motion. All of that stimulation keeps the prefrontal cortex (the thinking, reasoning, organizing part of the brain) engaged and somewhat in charge. The amygdala, which is the brain's primary threat-detection center, is still running in the background, but the busyness of the day tends to suppress its louder signals.

Then bedtime arrives. The lights go down. The house gets quiet. The screens go off. The distractions vanish. And the amygdala, which has been patiently queuing up every mildly stressful moment from the past fourteen hours, finally gets its turn.

This is not a malfunction. The amygdala is doing its job. Its evolutionary purpose is to scan for threats during quiet, low-stimulation periods, because historically that is when danger was most likely to approach. It does not know the difference between a predator and a friendship problem. It responds to both with the same alerting chemistry: a small flood of cortisol and norepinephrine, a racing heartbeat, a mind that suddenly cannot stop producing thoughts.

Diagram 10.1: The Bedtime Amygdala Activation

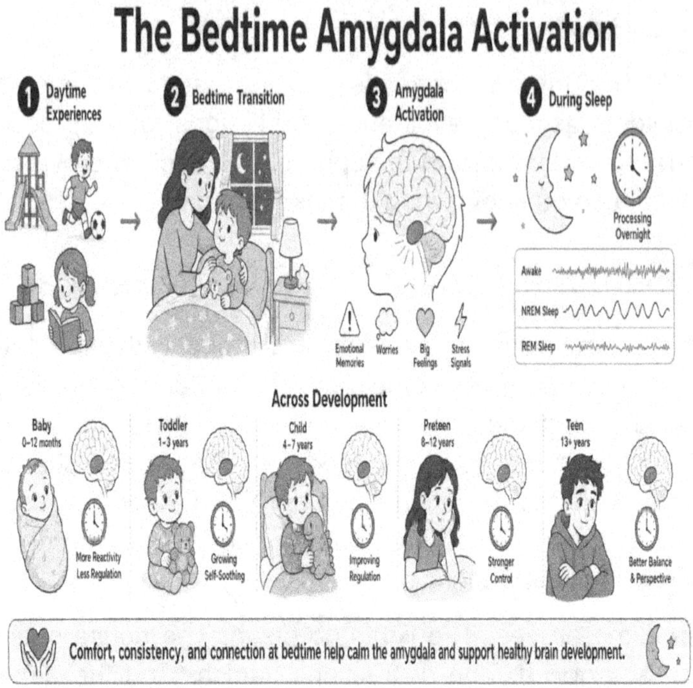

For children whose baseline anxiety is higher, this activation is more intense and more sustained. But even children who are not clinically anxious can experience this phenomenon after a hard day, a stressful week, or a big transition. The amygdala does not discriminate between sensitive kids and resilient kids. It activates in all of them. The difference is how quickly the prefrontal cortex can step back in and reassure the system that the threat is not real.

Here is the important part for parents: you cannot simply tell an activated amygdala to calm down. Rational conversation, firm boundary-setting, and reassurance delivered in an agitated voice do not reach the amygdala. The amygdala responds to the nervous system state of the person it is with. More on that shortly.

The Default Mode Network at Night

There is a second neurological actor in this story: the default mode network. This is a set of brain regions that activate when the brain is not focused on an external task. You have experienced it yourself. It is the mental state of lying in bed thinking about everything you said in a meeting last Tuesday, or running through tomorrow's to-do list, or replaying a difficult conversation. The default mode network is sometimes called the "resting state" network, but it is not restful at all. It is the brain's internal chatter, and for anxious people, it tends to produce anxious content.

In children, the default mode network matures through late childhood and into adolescence. This is actually part of why middle schoolers and teenagers struggle so much with nighttime rumination. Their default mode networks are becoming more sophisticated and more active, giving them a richer capacity for self-reflection and, when that capacity is turned up too high, a richer capacity for self-criticism and worry.

Diagram 10.2: Default Mode Network at Bedtime

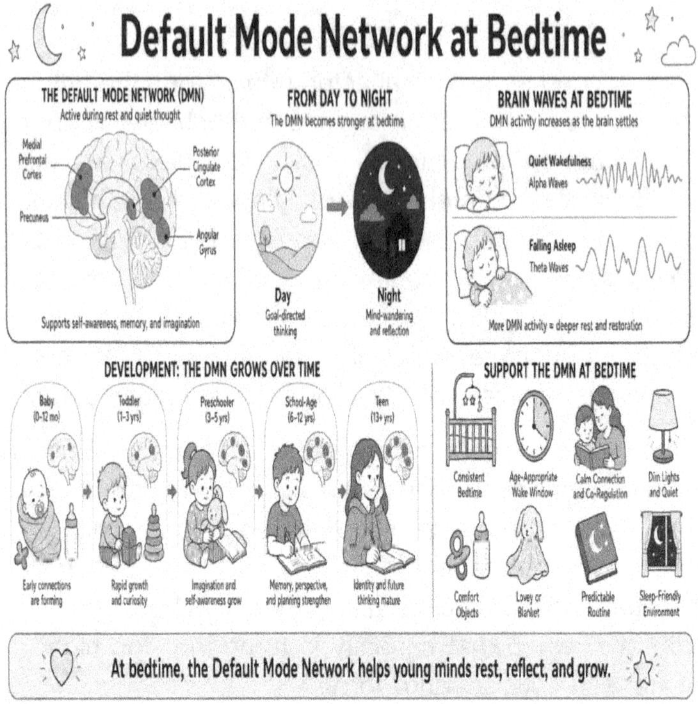

For younger children, the default mode network produces simpler anxiety content: fears about separation, fears about darkness, fears about things under the bed. For older children and adolescents, the content becomes more abstract and more future-oriented: social, academic, and existential fears. But in both cases, the underlying mechanism is the same. The brain, unoccupied by external input, turns inward and begins to review, anticipate, and catastrophize.

Understanding this helps enormously. When your child says "I can't stop thinking," they are describing a real neurological experience, not a performance or a manipulation. Their brain has shifted into default mode, and the content it generates is frightening to them. They want it to stop just as much as you want bedtime to be over.

Developmental Fears by Age

Anxiety at bedtime is not one thing. It changes shape as children grow, and recognizing the shape that matches your child's age helps you respond in a way that actually fits their developmental stage rather than working against it.

Infants and young toddlers (zero to eighteen months) experience separation anxiety as their primary bedtime challenge. This is not cognitive worry in the adult sense. It is a primal response to the absence of the attachment figure, which the developing brain genuinely interprets as a survival threat. Every responsive reunion teaches the brain that separation is temporary and safe. No amount of reasoning helps here, because language and reasoning are not yet online. Presence, warmth, and predictable return are the tools that work.

Diagram 10.3: Developmental Fears by Age

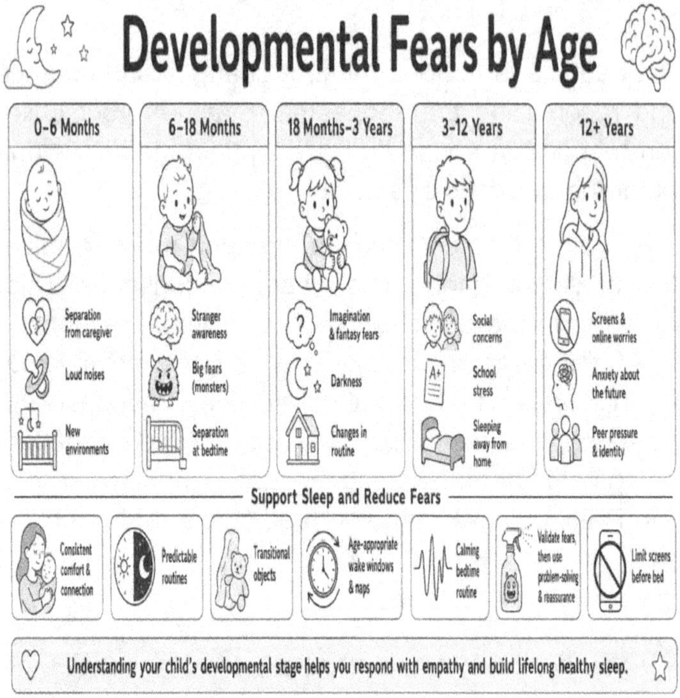

Toddlers and preschoolers (eighteen months to five years) add imagination to the mix, and imagination at this age produces monsters, shadows, and things in closets. This is actually a developmental achievement: they are now sophisticated enough to generate mental images that feel real. The mistake is trying to disprove the monster with logic ("monsters aren't real, sweetie"). Their brain is not yet equipped to use abstract logic to override a felt experience. What works better is entering the imaginative frame with them: checking together, leaving a small nightlight on, creating a "monster spray" (water in a spray bottle), and honoring their fear as real to them while helping them feel safer.

Children from about five to eight years begin to encounter existential fears: a grandparent getting sick, awareness of death, worry about family safety, concern about fires or disasters. These fears are signs of cognitive development, specifically the new ability to project into the future and understand that bad things can happen. They are heavy to carry in a small body, and they need something different than dismissal or over-reassurance. Sitting with the fear for a moment before offering comfort works better: "Yeah, that is a scary thing to think about. You know what else is true? Here's who keeps us safe..."

School-age children (roughly eight to twelve) add social and academic anxiety: friendship conflicts, fear of failure, worry about performance. This is when worry content most closely resembles adult worry, and this is the age group that responds best to the cognitive-behavioral techniques in this chapter. They have enough language and self-awareness to learn named techniques and apply them with practice.

Teenagers, unsurprisingly, carry the most complex anxiety loads: identity, relationships, academic pressure, world events, and the internal storm of a developing sense of self. Teen bedtime anxiety often looks like extreme lateness in falling asleep, phone dependency, and resistance to any technique that feels "babyish." Approaches that work for this age group tend to emphasize autonomy: teaching the skill and letting the teen decide how and whether to apply it.

CBT-I Principles Adapted for Children

Cognitive Behavioral Therapy for Insomnia, commonly called CBT-I, is the gold-standard treatment for insomnia in adults. It was developed for adult brains and adult reasoning capacities, but its core principles translate remarkably well to children when adapted thoughtfully. Three of its key components are particularly relevant here: stimulus control, sleep compression, and cognitive restructuring. None of them are complicated, and none require a clinical degree to apply at home.

Before diving into the techniques, it helps to understand what CBT-I is actually doing at the brain level. The core insight is that insomnia (whether anxiety-driven or not) is partly a learned problem. The brain learns to associate the bed with wakefulness, vigilance, and worry. Once that association is formed, lying down in bed reliably triggers arousal instead of drowsiness. The goal of CBT-I is to break that learned association and replace it with a new one. The techniques below all work toward the same aim: making the bed a cue for safety and sleep, not a cue for dread.

Stimulus control is the idea that the bed should be associated exclusively with sleep, not with worry, screens, or stimulating activity. For children, this means that if a child is lying in bed worrying for more than about twenty minutes, it is actually more effective to get up and sit in a calm, dimly lit spot for a few minutes before returning to bed. This interrupts the learned association between "lying in bed" and "feeling anxious." Over time, the bed becomes a cue for sleep rather than a cue for worry. The same principle applies to avoiding screens in bed and keeping homework, arguments, and intense conversations out of the bedroom entirely.

Diagram 10.4: CBT-I Adapted for Kids

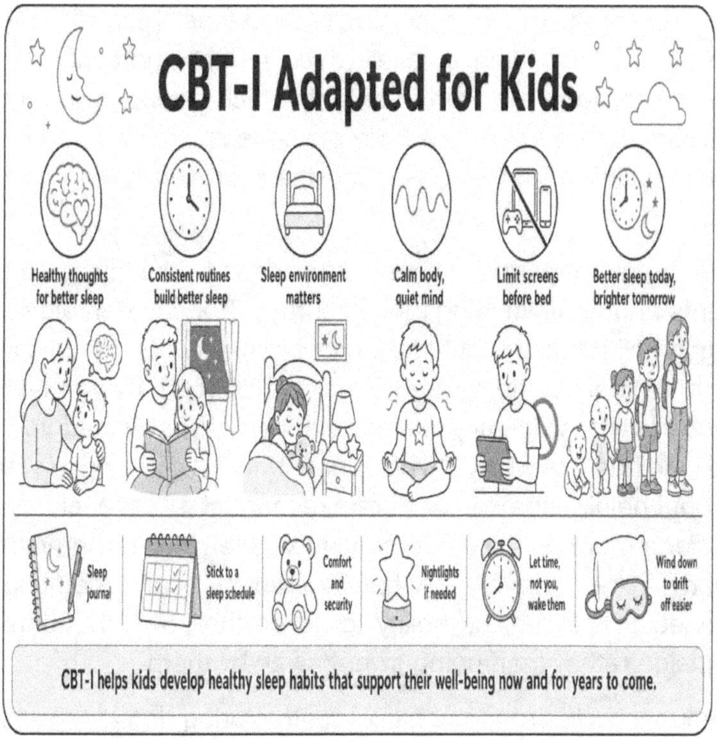

Sleep restriction, in its adult form, involves limiting time in bed to match actual sleep time, which quickly consolidates sleep and reduces anxiety from lying awake for extended periods. In the child-safe version, this is not about severely limiting sleep. Instead, it means being thoughtful about not sending children to bed much earlier than they are genuinely sleepy, even when you desperately want them asleep. A child sent to bed at 7:30 when their natural sleep drive does not build until 8:30 will spend an hour lying awake, and that hour will be filled with anxious thoughts. Matching bedtime more closely to when the child is actually sleepy, while keeping wake time consistent, tends to significantly reduce sleep-onset anxiety.

There is a particular trap that well-meaning parents often fall into: making bedtime earlier in response to a child's sleep problems, hoping that more time in bed will translate into more sleep. For anxious children, this almost always backfires. More time in bed means more time lying awake, which means more time for the amygdala to generate worry content, which deepens the learned association between bed and anxiety. A temporarily later bedtime, counterintuitive as it feels, is often the more effective short-term move.

Cognitive restructuring is the practice of identifying anxious thoughts and gently testing their accuracy. For children old enough to engage with this (roughly seven and up), you can teach a simple version: "What is the worry? What's the worst thing that could happen? What is actually likely to happen? What would you tell a friend who had this worry?" This is not about invalidating the feeling. It is about helping the thinking brain reconnect with the feeling brain and offer a more balanced perspective. When children practice this regularly, they develop what therapists call metacognitive awareness, meaning the ability to notice their own thoughts from the outside rather than being swept away by them.

For younger children, cognitive restructuring looks less like a structured conversation and more like narrative. You sit with the child and help them tell the worry story with a different ending: "The monster is actually grumpy because it wants a friend. What if we drew it a picture to make it feel better?" You are not dismissing the fear. You are modeling the skill of holding a scary thought and gently turning it toward something more manageable. Over years of practice, this becomes internalized. The child who once needed you to rewrite the monster story learns eventually to do it for themselves.

The Worry-Time Technique

One of the most reliably effective tools for bedtime anxiety in children is almost embarrassingly simple: move the worry earlier. This is called worry time or the worry window, and the research behind it is solid. The core principle is that worries do not disappear when you ignore them. They wait. And where they tend to wait is bedtime, when there is finally no competition for attention.

Worry time is a designated ten-to-fifteen-minute slot earlier in the evening (after school but at least two hours before bed, typically during dinner cleanup or a quiet moment around 6:30 or 7:00 p.m.) during which the child is actively invited to worry. Not just worry silently, but to speak worries aloud, write them down, or draw them. The parent's role is to listen without immediately solving the problem, to take the worry seriously, and to help the child write a very short "plan" for any worry that has an actual solution.

Diagram 10.5: The Worry-Time Technique

The script is simple. "Okay, it is worry time. What is on your mind? Let's get it all out." You listen. You validate. You do not minimize ("Oh, that's nothing to worry about") or catastrophize ("Oh no, that does sound terrible"). You help them sort their worries into two buckets: those they can do something about and those they cannot control. For the first bucket, you make a tiny plan together. For the second bucket, you practice acknowledging the worry without solving it: "Yeah, that is not something we can control. It is okay to notice it and then let it be."

Then, when worry shows up at bedtime, both you and your child have a shared script: "We already did worry time tonight. This worry has already been heard. It doesn't need to be solved again right now. Let's put it in the jar." (More on the jar shortly.) This works because it is not dismissive. The worry was genuinely attended to. The brain got its attention. The bedtime recurrence is simply a habit, and habits can be redirected.

For parents of younger children (three to six), worry time can be as simple as a bedtime question over a snack: "Any hard feelings we should talk about before bath time?" The formality of the technique matters less than the consistency. What the brain learns is: feelings get attended to before bedtime, not during.

Belly Breathing and the Parasympathetic Reset

When a child is lying in bed with a racing heart and a mind that won't quit, their sympathetic nervous system has activated. Talking them out of it does not work because language lives in the cortex, and the stress response is subcortical. What works is activating the parasympathetic nervous system throughout the body, and the most accessible on-ramp is the breath.

Slow, extended exhalation activates the vagus nerve, which is the primary driver of the parasympathetic "rest and digest" response. When the exhale is longer than the inhale, heart rate slows, cortisol begins to drop, and the amygdala receives a genuine neurochemical message that the threat has passed. Children can be taught this, and many find it genuinely powerful once they understand why it works.

Diagram 10.6: Belly Breathing Steps

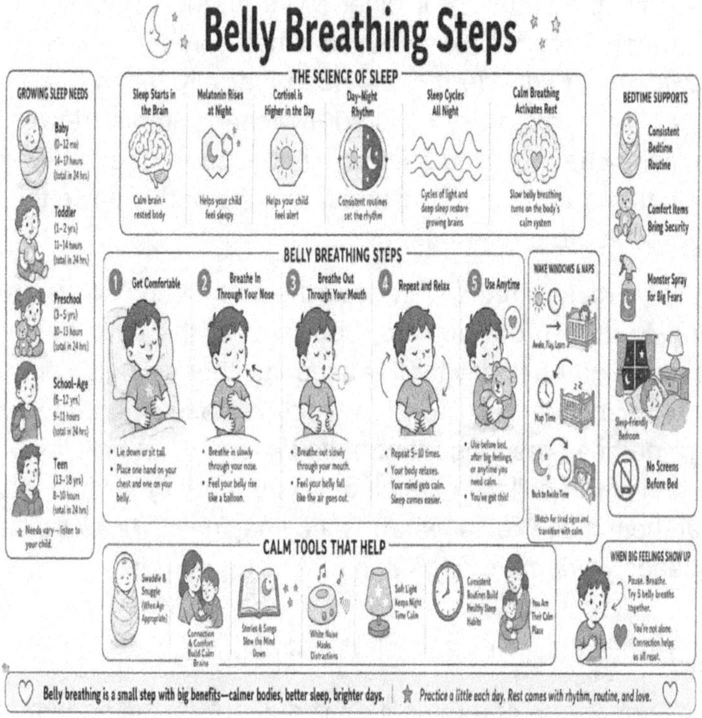

For children under six, the easiest version is "smell the flower, blow out the candle." Breathe in slowly through the nose (three counts, imagining smelling a flower), then breathe out slowly through the mouth (five counts, imagining blowing out a birthday candle). Three to five repetitions are enough to shift the nervous system noticeably.

For children six and up, box breathing works well: breathe in for four counts, hold for four counts, breathe out for four counts, hold for four counts. The "box" is easy to visualize, and the rhythm is predictable enough to give the mind something concrete to focus on, which interrupts the default mode network's anxious content.

Body scans are a companion practice. Guide your child to slowly bring their attention to each part of the body from toes to head, noticing any tension without trying to fix it. For younger children, you can frame this as "checking in with your body, like a superhero scanning for problems." For older children, the instruction can be more straightforward: "Just notice what you feel in each part, without trying to change it." The scan typically takes three to five minutes. It leaves most children noticeably more relaxed, not because they have solved anything, but because they have moved attention from the mind to the body, and the body, unlike the mind, tends to be in the present moment.

Progressive muscle relaxation takes the body scan one step further. Working from the feet upward, you instruct the child to squeeze each muscle group tightly for five seconds, then release completely. The contrast between tension and release triggers a deep parasympathetic response that goes beyond what breathing alone can achieve. For children, framing this as "make your feet into fists, then let them melt like snow" adds the imaginative layer that makes body-based work more engaging for this age group.

Visualization Scripts for Children

Visualization is one of the most powerful tools in the bedtime toolkit and one of the most enjoyable for children with vivid imaginations. The purpose is not distraction exactly. It is redirection: giving the default mode network something rich and pleasant to generate instead of anxious content. Two specific scripts are worth knowing in detail, and both can be started tonight.

The brain does not fully distinguish between imagined and real experiences in its emotional and physiological responses. When your child imagines a warm, safe, sensory-rich place with enough detail, their nervous system responds as if that place is real: heart rate slows, breathing deepens, muscle tension decreases. This is not a metaphor. Neuroimaging research shows that vivid mental imagery activates many of the same brain regions as actual perception. You are not just entertaining your child at bedtime. You are offering their nervous system a genuine physiological experience of safety.

The Safe Place Visualization asks the child to imagine a place where they feel completely calm and safe. It can be real (a grandparent's kitchen, a beach they visited, a cozy corner of their own room) or imaginary (a treehouse in a magical forest, a spaceship, an underwater cave with friendly fish). Once the child has a location, you guide them through sensory details: "What does it smell like there? What can you hear? Is it warm or cool? What can you see when you look around?" The richer the sensory detail, the more thoroughly the brain's imagery system engages, and the less bandwidth is available for anxious thought. After several nights of practice, the safe place becomes a reliable mental destination the child can visit on their own.

The Worry Jar (or Worry Box) Method is a combination of visualization and physical ritual. During the day, worries are written or drawn on small slips of paper and placed into a decorated jar or box. Before bed, the child symbolically closes the jar and places it on a shelf. The script is: "All your worries are safe in the jar. They don't need to be thought about again tonight. The jar is keeping them for you until morning." For children too young to write, worries can be drawn, or a parent can write them down as the child dictates. What makes this work is the physical act of externalization. The worry moves from inside the mind to outside the body, and the brain genuinely responds to that shift. It is not magic; it is a concrete way of helping the child's nervous system practice the sensation of setting something down.

Diagram 10.7: The Worry Jar Method

For children who resist the jar or find it too babyish, a worry journal achieves the same end with a more grown-up feel. A notebook on the nightstand for "tonight's thoughts" lets older children and adolescents dump worries onto the page before bed. The act of writing externalizes and partially processes the worry, reducing its grip on the resting mind. Some children naturally evolve from the jar to the journal as they grow, which is a lovely arc to watch.

When Anxiety Crosses a Clinical Line

Most childhood bedtime anxiety is normal, variable, and responsive to the tools in this chapter. But some anxiety is bigger than lifestyle techniques can address, and it is important to know the signs.

Seek support from your child's pediatrician or a licensed child therapist when you notice any of the following patterns persisting for more than two to four weeks: the anxiety is significantly impacting sleep every night, not just on hard nights; the child is avoiding school, activities, or social situations in addition to resisting bedtime; physical symptoms such as stomachaches, headaches, or nausea appear regularly at bedtime or in anticipation of separation; the child expresses persistent feelings of hopelessness or low self-worth; panic-like symptoms (heart racing, trouble breathing, feeling of doom) are occurring at bedtime; or the anxiety has intensified after a trauma, loss, or significant family disruption.

Diagram 10.8: When to See a Pediatric Therapist

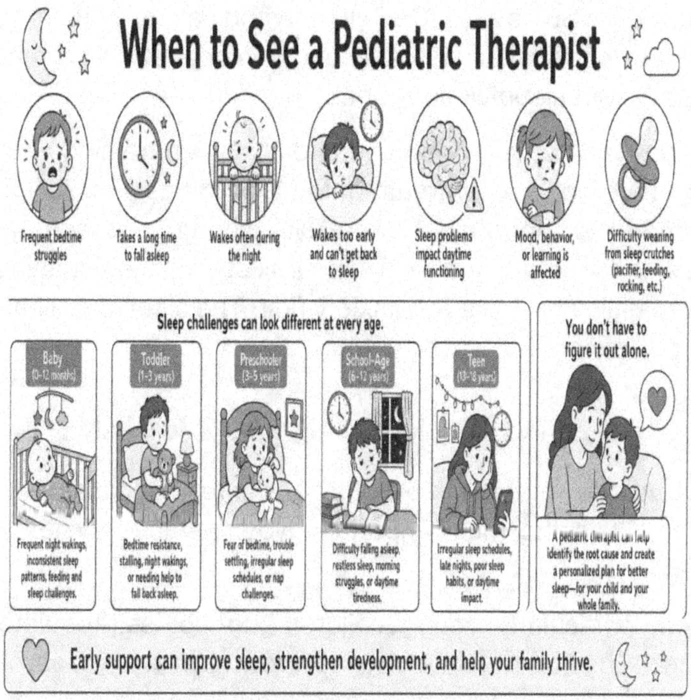

Anxiety disorders are among the most common mental health conditions in childhood, affecting roughly one in five children at some point. They are also among the most treatable. Cognitive Behavioral Therapy for childhood anxiety, delivered by a trained child therapist, has one of the strongest evidence bases of any psychological treatment for children. Early intervention changes trajectories. There is nothing to feel guilty about in seeking help, and quite a lot to feel proud of in recognizing that your child needs and deserves it.

A note on medication: for moderate to severe childhood anxiety, medication (typically an SSRI) is sometimes recommended alongside therapy. This decision belongs with your pediatrician and, when appropriate, a child psychiatrist. It is not a decision you need to make today, but it is worth knowing that the option exists and is supported by genuine evidence. Many children do very well with therapy alone. Some do better with a combination. What matters is that the anxiety is taken seriously and treated, not simply managed indefinitely with sleep workarounds.

It is also worth naming one thing directly: sleep training and sleep protocols are not anxiety treatment. If a child has a clinical anxiety disorder and you apply a behavioral sleep program to them, you are addressing the wrong problem. The anxiety will persist and so will the sleep difficulty. Treat the anxiety first. The sleep often improves as the anxiety comes down.

Co-Regulating with an Anxious Child

Here is the part of this chapter that is hardest for most parents to hear, and also the most important: your nervous system is the intervention.

Children, particularly young children, do not self-regulate their emotions. They co-regulate. They borrow regulatory capacity from the calm nervous system of a trusted adult, the way a cold room borrows heat from a warm one. When you sit on the edge of the bed with a tight jaw and a flat voice, trying to project calm while your own cortisol is elevated from thirty-seven hard minutes of getting everyone to bed, your child's amygdala does not hear your words. It reads your physiology. And it escalates to match.

This is not blame. This is neuroscience. And knowing it is actually liberating, because it shifts the question from "what should I say to my anxious child" to "what can I do with my own body right now." The answer is often: slow your own exhale, soften your face, lower your voice, drop your shoulders. You do not need to achieve total serenity. You need to shift a few degrees toward regulated. Children are extremely sensitive to even small changes in a parent's physiological state.

Diagram 10.9: Parent Co-Regulation Map

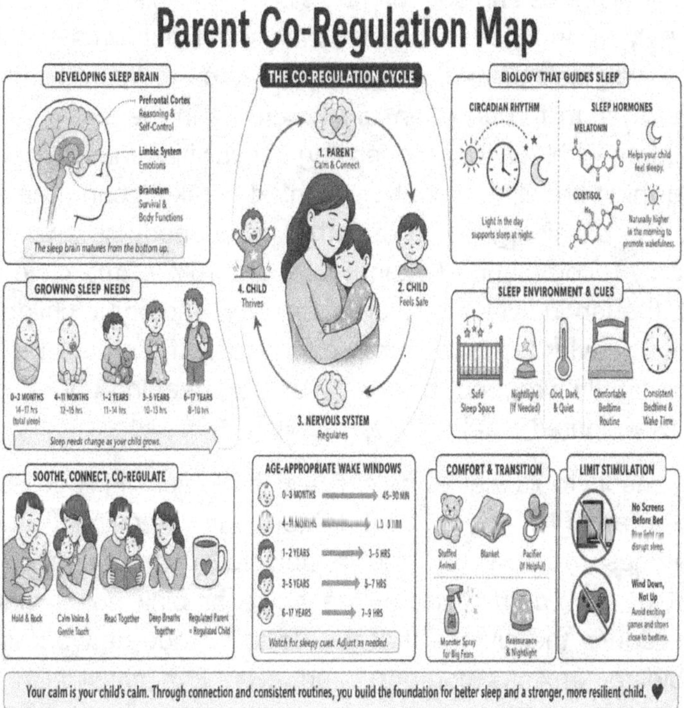

A 60-second parent reset before entering an anxious child's room can genuinely change what happens next. It can be as simple as standing outside the bedroom door, doing three slow breaths, consciously dropping physical tension from the shoulders and jaw, and saying silently to yourself: "I can handle this. My child is safe.

This is not an emergency." That is not false reassurance. It is a recalibration that reaches your amygdala in the only language it speaks: body sensation.

When you enter the room already a few degrees calmer, your child's nervous system immediately picks up on that signal. They may not settle instantly. But the window for settling opens. The escalation cycle slows. You become what child development researchers call a "co-regulating presence," and that presence is genuinely one of the most powerful therapeutic tools.

Weighted blankets and sensory tools belong in this section because they work through a similar mechanism. Deep-pressure touch, which a weighted blanket provides, activates the parasympathetic nervous system through sensory receptors in the skin and joints. For children with sensory processing sensitivities, autism spectrum traits, or high baseline anxiety, weighted blankets can dramatically reduce the physiological arousal that drives bedtime anxiety. The commonly recommended weight is about 10% of the child's body weight, though individual preferences vary, and some children find lighter or heavier weights more comfortable. If your child is drawn to being tightly wrapped, heavy covers, or sleeping under multiple layers, a weighted blanket is worth trying.

Other sensory tools that some children find helpful include lavender aromatherapy (there is modest evidence for its calming effect on the nervous system), soft tactile objects such as a silky-edged blanket or a smooth stone, white noise or pink noise machines (which reduce the silence that the default mode network finds so available for anxious content), and even firm physical pressure delivered by a parent through a long, slow back-squeeze at goodnight.

None of these tools is a cure. All of them are legitimate physiological levers. Used consistently as part of a calming bedtime routine, they shift the nervous system's baseline and make the other cognitive tools- the worry time, the breathing, the visualization- land more easily.

Building a Worry-Ready Home

The techniques in this chapter work best when they are part of a broader family culture around emotional honesty and calm. This does not mean your family needs to be perfect or serene. It means that worry is named, normalized, and regularly attended to, not just at bedtime, when everyone is exhausted. A bedtime anxiety tool that a child has never practiced in the daylight will not land well at 9:00 p.m. when the feelings are most intense. The tools need to be familiar and associated with positive experiences before they are called upon in the hard moments.

A few practices that support this over time: making it normal to name feelings at dinner ("one hard thing from today and one good thing" as a simple family ritual); modeling your own emotional regulation aloud so children see what managing anxiety actually looks like ("I'm feeling a little stressed about tomorrow. I'm going to take three slow breaths. Want to do it with me?"); and avoiding the instinct to immediately reassure or fix, which can inadvertently send the message that the worry was too big to sit with. When you reassure too quickly, you short-circuit the process of learning that worry is survivable. Sometimes the most powerful response to a child's worry is a quiet acknowledgment: "Yeah, that is hard to sit with. I'm right here."

It is also worth saying something directly about screens at bedtime, because this conversation comes up in almost every family dealing with childhood anxiety. Screens are not just a sleep disruptor because of blue light. For anxious children especially, they serve as an anxiety management tool. The phone, the tablet, the TV show-all of them provide enough stimulation to keep the default mode network occupied. When you remove the screen, the worries rush in. This is why many anxious children and teenagers cling to devices at bedtime with what feels like desperation. They are not being defiant. They are self-medicating. The solution is not simply to remove the screen (though reducing blue-light exposure matters) but to replace it with something that serves the same regulatory purpose: a warm conversation, a familiar audiobook, a slow breathing practice. The goal is to gradually shift from external stimulation as the anxiety manager to internal skills as the anxiety manager, and that shift takes time and patience.

One more thing worth naming, for parents who are reading this chapter while recognizing themselves in the anxious child: anxiety has a significant genetic component, and many parents of anxious children are themselves more anxious than average. That is not a flaw. It is a biological reality that you did not choose and did not cause. What it means practically is that you may find some of the tools in this chapter useful for your own nervous system as well as your child's. The belly breathing works for adults. The worry-time technique works for adults. The sixty-second reset works for adults. If you find yourself lying awake at night running your own version of the bedtime amygdala activation, you have complete permission to use everything in this chapter for yourself.

Children who grow up in families where worry is openly acknowledged and gently attended to tend to develop stronger anxiety regulation over time. They internalize the message that difficult feelings are survivable, that the people who love them will not be overwhelmed by their distress, and that they have real tools available when their minds go dark. These are not small things. They are the building blocks of emotional resilience, and they start with bedtime.

Try This Tonight

For this age range: 5 to 12 years

1. At dinner cleanup or bath time tonight, introduce a five-minute "worry window." Ask your child one open question: "Anything on your mind we should talk about before bed?" Listen more than you respond.

2. Together, write down any worries that come up on a small slip of paper. Do not try to solve them all. Just acknowledge

each one and confirm you heard it. For any worry that has a practical next step, write one short plan.

3. Place the slips of paper in a jar, box, or drawer. Say together: "These are safe here. They don't need to be thought about again tonight."

4. Build in two to three minutes of belly breathing before lights out. Use "smell the flower, blow out the candle" for younger children (in for three, out for five) or box breathing for older children (in for four, hold for four, out for four, hold for four).

5. Before entering your child's room if they call out anxiously, take thirty seconds to slow your own breath, drop tension from your jaw and shoulders, and soften your face. Your nervous system is the first tool.

6. If your child is still awake and worrying after twenty minutes, invite them to sit quietly in a dimly lit spot for five to ten minutes before returning to bed. This is the stimulus-control principle: breaking the association between lying in bed and feeling anxious.

7. Keep this routine consistent for at least two weeks before judging its effectiveness. Anxious habits built over months do not shift overnight, but most children show meaningful improvement within 1 to 2 weeks of consistent practice.

Parent Takeaways

- Bedtime anxiety is neurologically predictable, not a behavior problem. The brain's threat-detection system activates when stimulation fades, and the default mode network generates worry content. Understanding this changes how you respond.

- Move worry processing earlier in the evening. Worry that is addressed at 6:30 p.m. is less likely to resurface at 9:00 p.m. Worry time, the worry jar, and a brief feelings check-in at dinner all achieve this.

- Body-based tools reach the nervous system in ways that words cannot. Belly breathing, progressive muscle relaxation, weighted blankets, and physical reassurance all work through physiological pathways. Use them alongside, not instead of, conversation.

- Your calm is the most powerful tool in the room. Children co-regulate through their caregivers' nervous systems. A sixty-second pre-bedtime reset for yourself is not indulgent; it is clinical.

- Persistent anxiety that impacts daily functioning warrants professional support. CBT for childhood anxiety is highly effective. Seeking help is an act of good parenting, not a sign of failure.

11 Troubleshooting and Red Flags

You have done everything right. You read the books, set the routines, dimmed the lights at seven-thirty, kept the room cool and dark, and never wavered on the schedule. And yet your child still looks exhausted every morning. Still snores through the night. Still comes home from school so wired you wonder how anyone could call it tiredness, but then crashes without warning and is impossible to rouse. You are past the stage of wondering whether you are doing something wrong at bedtime. Something else is going on. You can feel it.

This chapter is for that moment. It is for the parent who has already tried the behavioral fixes and knows in their gut that the problem runs deeper. It is also for the parent who is only now noticing patterns that were always there, ones that blended into the general chaos of raising young children. Snoring that you assumed was just congestion. Nighttime episodes that scared you but felt too strange to explain easily. A child who cannot stay awake through a movie but cannot fall asleep at bedtime either. These patterns have names, explanations grounded in pediatric sleep science, and, for most, effective treatments.

Pediatric sleep disorders are more common than most parents realize, and they are genuinely under-recognized. That is partly because the symptoms often look like something else entirely: behavioral problems, attention difficulties, mood swings, hyperactivity, or just the normal chaos of childhood. A child with obstructive sleep apnea does not necessarily look as sleepy as an adult would. They may look wired, defiant, and unfocused. A child with narcolepsy might seem lazy or emotionally volatile before anyone connects the dots. A child with restless legs syndrome might seem like a kid who cannot sit still. The medical and behavioral territories overlap, and it takes a deliberate diagnostic lens to see what is really happening.

So this chapter is about building that lens. Not to alarm you. Not to send you spiraling through every possible diagnosis for your child. But to give you a clear framework for knowing when a sleep problem is within the range of normal behavioral and developmental variation, and when it is something that genuinely needs a medical conversation. You deserve that framework. Your child deserves it even more.

Chapter Summary

This chapter walks through the most important pediatric sleep disorders and red-flag symptom patterns, providing a practical way to distinguish behavioral from medical sleep challenges. It covers pediatric obstructive sleep apnea in depth, including why it looks different in children than in adults and why adenotonsillectomy is often the first and most effective treatment. It addresses restless legs syndrome and its connection to iron levels, pediatric narcolepsy and the rare but real phenomenon of cataplexy, and the wide world of parasomnias, from night terrors and sleepwalking to confusional arousals and the rare REM behavior disorder.

The chapter also covers bedwetting, bruxism, sleep-related anxiety, and pediatric melatonin use, giving you a grounded, science-based view of what is appropriate, what to avoid, and when to stop managing things at home and start a conversation with your pediatrician or a pediatric sleep specialist. By the end, you will know what a pediatric sleep study involves, what questions to ask at a medical visit, and what the road to diagnosis and treatment actually looks like.

What This Chapter Will Cover

- Pediatric obstructive sleep apnea. Snoring and restless sleep are the most common signs, but daytime presentation in children often resembles hyperactivity rather than sleepiness. This section explains why OSA works differently in children than adults, what the warning signs look like at every age, and why the first treatment is usually a surgical referral rather than a CPAP machine.

- Restless legs syndrome in children. RLS in kids is frequently missed because children describe the sensation differently than adults do, and the urge-to-move pattern is often mistaken for behavioral restlessness. This section explains the connection to ferritin and what a pediatrician should check.

- Pediatric narcolepsy. Narcolepsy is rare but real, and it is chronically underdiagnosed in children. Cataplexy, sleep attacks, and the full symptom picture are explained clearly so parents know what to watch for and when an urgent referral is warranted.

- Parasomnias. Night terrors, sleepwalking, confusional arousals, and REM behavior disorder each arise from different sleep stages, have different risk profiles, and require different responses. This section gives you a clear map of each, including when to be concerned and when to wait it out.

- Bedwetting, bruxism, and sleep-related anxiety. These three common childhood issues are often underestimated as sleep disruptors. The section on each explains the difference between what is developmentally normal and what warrants evaluation.

- When to see a pediatric sleep specialist. This section gives you a concrete, practical checklist so you do not have to guess whether the problem is big enough for a specialist.

- What a pediatric sleep study involves. Most parents have no idea what actually happens during a polysomnogram. This section walks through the process from arrival to report so you know what to expect and how to prepare your child.

- Medications and melatonin in children. Pediatric melatonin use has exploded in recent years. This section explains what the research actually says, what the AAP recommends, and why starting with behavioral solutions is still the right first move before opening the supplement cabinet.

Pediatric Obstructive Sleep Apnea

Obstructive sleep apnea in children is not a smaller version of adult OSA. It is a meaningfully different condition with distinct causes, symptoms, and first-line treatment. Understanding those differences is one of the most important takeaways for a parent from this chapter.

In adults, OSA is most often caused by excess weight and the resulting collapse of soft tissue in the throat during sleep. In children, especially those between two and eight years old, the most common anatomical culprit is enlargement of the tonsils and adenoids. Because the airway in children is proportionally much smaller than in adults, even mildly enlarged lymphoid tissue can cause significant airway obstruction during sleep. The obstruction does not have to be complete to cause problems. Partial obstruction, in which breathing becomes labored and inefficient without stopping entirely, is enough to fragment sleep, reduce oxygen saturation, and repeatedly activate the stress response throughout the night.

Diagram 11.1: Pediatric OSA Warning Signs

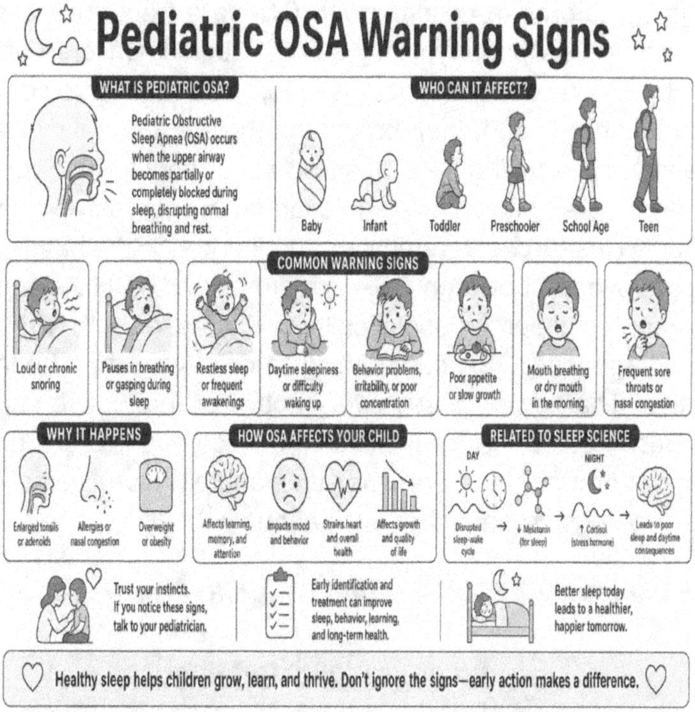

The most recognizable sign of pediatric OSA is habitual snoring, meaning snoring that occurs most nights rather than only during a cold or allergy season. If your child snores regularly, that alone is worth mentioning to your pediatrician. But snoring is not the only warning sign, and in many children it is accompanied by a cluster of other observations that together make the picture clear. You might notice your child sleeping with their mouth open, even when they are not congested. You might notice that they breathe loudly during sleep, or that they seem to be working hard to breathe, with visible chest or abdominal movement. In more significant cases, you might witness brief pauses in breathing followed by a gasp, snort, or sudden movement. Those witnessed apneas are a direct red flag, and a child whose breathing regularly pauses during sleep should be seen by a pediatrician promptly.

Daytime symptoms in children with OSA are frequently misread. Adults with sleep apnea typically feel exhausted. Children often do not exhibit obvious fatigue in response to chronic sleep fragmentation. Instead, their brains upregulate arousal systems to compensate, resulting in hyperactivity, impulsivity, difficulty concentrating, emotional dysregulation, and aggressive or defiant behavior. Researchers studying pediatric OSA have found significant overlap between OSA symptoms and ADHD symptoms, and a meaningful percentage of children diagnosed with ADHD may have undiagnosed OSA contributing to or driving their behavioral difficulties. That is not to say every active or inattentive child has OSA. But if the behavioral picture is accompanied by habitual snoring or unrefreshing sleep, the sleep angle is worth investigating before or alongside any other evaluation.

Diagram 11.2: OSA: Adult vs. Child Presentation

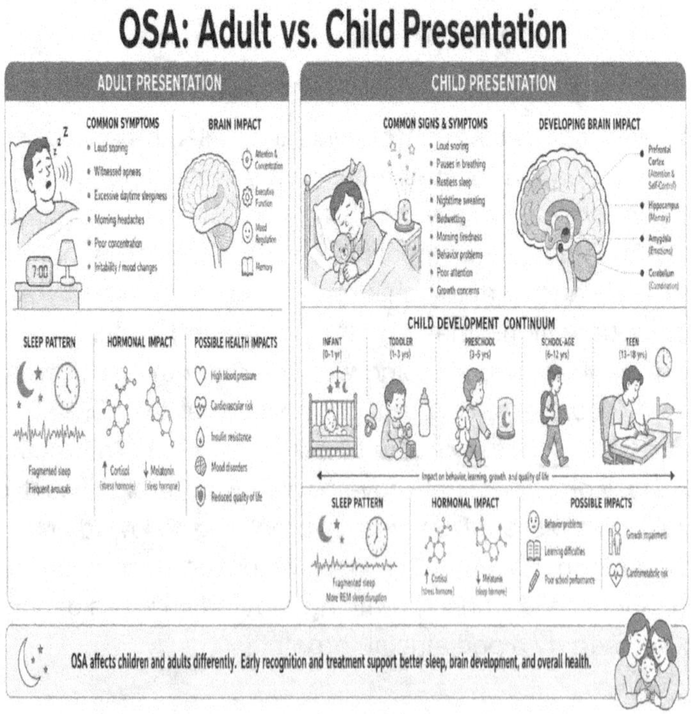

Children with OSA may also be difficult to wake in the morning, seem groggy or grumpy for an extended period after waking, and fall asleep readily in the car or during quiet activities despite what should be an adequate total sleep time. Bedwetting past the expected age of continence can also be a sign, because elevated carbon dioxide from impaired breathing during sleep increases urine output. Mouth breathing during the day, chronic nasal congestion, and a history of frequent ear infections round out a clinical picture that pediatricians and ENT specialists are trained to recognize.

The first-line treatment for pediatric OSA caused by adenotonsillar hypertrophy is adenotonsillectomy, the surgical removal of the adenoids and tonsils. This is meaningfully different from how adult OSA is managed, where weight loss, positional therapy, and CPAP are typically the initial approaches. In children, surgery often resolves OSA completely, particularly when the child is otherwise healthy and the tonsils and adenoids are genuinely enlarged. Studies have documented significant improvements not just in breathing during sleep but in behavior, attention, quality of life, and growth velocity following adenotonsillectomy in children with OSA. The surgery is not appropriate for every child, and the decision involves weighing the degree of obstruction, the child's overall health, and family considerations. But parents should know that it is a real, often curative option, not a last resort.

For children in whom adenotonsillectomy is not sufficient or not indicated, pediatric CPAP is available and effective. CPAP in young children requires careful fitting, behavioral acclimatization, and ongoing support, but it is used successfully in children as young as a few months old when necessary. Positional changes, weight management in older children, and intranasal corticosteroids for mild cases round out the pediatric treatment toolkit. Talk to your pediatrician if you see the warning signs. A referral to a pediatric ENT or a certified pediatric sleep specialist is the right next step when OSA is suspected.

Restless Legs Syndrome in Children

Restless legs syndrome is one of the most underdiagnosed sleep-related conditions in children, partly because kids rarely describe it the way textbooks describe it. Adults with RLS typically report an uncomfortable urge to move their legs, often accompanied by crawling, tingling, or burning sensations that worsen in the evening and are relieved by movement. Children, especially younger ones, tend to describe it more vaguely: their legs hurt, their legs feel weird, they need to kick, they cannot be still. A parent watching from the outside may see a child who twists in the sheets, hangs their legs over the edge of the bed, or kicks repeatedly before finally settling. Bedtime becomes prolonged and frustrating for everyone, and the child genuinely cannot explain why they cannot stop moving.

Diagram 11.3: RLS and the Ferritin Connection

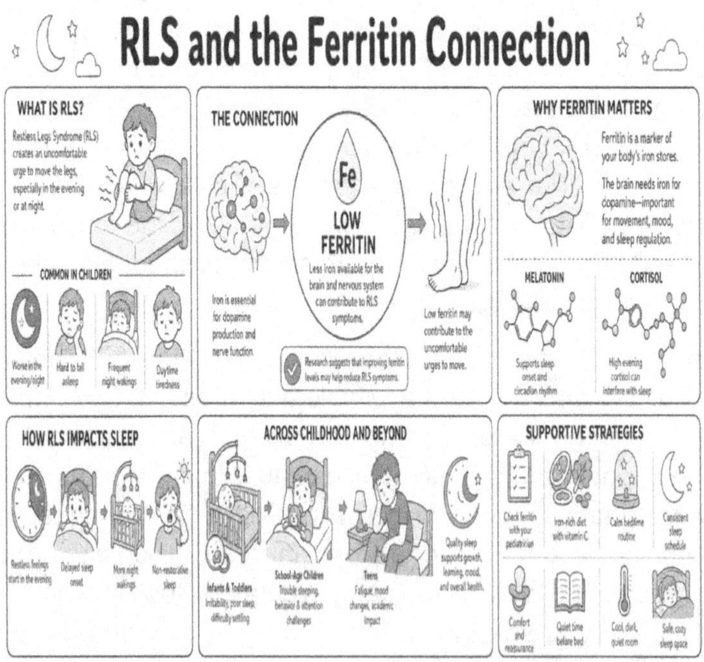

RLS in children has a strong and well-documented association with low ferritin, the protein that stores iron in the body. Unlike hemoglobin, which measures iron available in blood, ferritin reflects the brain's iron stores, and the central nervous system needs adequate iron to regulate dopamine pathways involved in motor control during sleep. Many children with RLS have ferritin levels that fall within the technically normal range but that are suboptimal for neurological function. Pediatric sleep researchers have found that ferritin levels below 50 micrograms per liter are associated with RLS symptoms, even when anemia is not present. A simple blood test can measure ferritin, and supplementing iron under medical supervision often produces dramatic improvement in RLS symptoms.

If your child shows the movement patterns described here, especially at bedtime, when they are lying still and trying to fall asleep, it is worth discussing with your pediatrician having their ferritin levels checked. This is a simple, low-burden first step that can make a significant difference. A physician should guide iron supplementation, since the appropriate dose and duration depend on the child's current levels and overall health. Do not supplement iron without first getting the blood test and talking to your pediatrician about what the numbers mean for your specific child.

RLS in children is also associated with a family history of the condition. If you or another close family member has RLS, the probability that a child in the family will have it increases substantially. There is also a relationship between RLS, ADHD, and the nighttime movement disorder called periodic limb movement disorder, which involves involuntary leg jerks during sleep. These conditions often co-occur and can compound each other's effects on sleep quality. A pediatric sleep evaluation can sort through the picture and identify the most helpful interventions.

Pediatric Narcolepsy

Narcolepsy is rare. That needs to be said first, because this section may prompt worry in parents of children who are just heavy sleepers or who seem tired frequently. But narcolepsy is also real; it is frequently missed for years before diagnosis, and the consequences of leaving it untreated are significant. Children with undiagnosed narcolepsy are often labeled as lazy, unmotivated, or emotionally unstable long before anyone connects their symptoms to a treatable neurological condition.

Narcolepsy is caused by the loss of hypocretin-producing neurons in the hypothalamus. Hypocretin, also called orexin, is a neuropeptide that stabilizes the transitions between sleep and wakefulness. Without it, the boundaries between sleep stages become fluid and unpredictable. People with narcolepsy can fall asleep suddenly during activities that would normally keep anyone awake, experience dream-like hallucinations as they are falling asleep or waking up, feel a brief paralysis upon waking, and, in the most recognizable form of the condition, experience cataplexy.

Diagram 11.4: Pediatric Narcolepsy Red Flags

Cataplexy is a sudden, brief loss of muscle tone triggered by strong emotion. Laughter is the most common trigger, but surprise, excitement, and even frustration can provoke it. In adults, cataplexy typically presents as sudden weakness or buckling of the knees, facial sagging, or head drop. In children, it often looks different and can be subtler, manifesting as a brief facial droop, a stumble, a momentary slackening of posture, or a sudden expressionless look during laughter. Very young children may show a grimacing or partial facial weakness that is easy to miss or attribute to something else. Children with cataplexy often intuitively suppress strong emotion to avoid triggering it, which can make them appear emotionally flat or overly controlled.

The other hallmark symptoms of narcolepsy are excessive daytime sleepiness (EDS), hypnagogic and hypnopompic hallucinations, and sleep paralysis. EDS in children with narcolepsy is profound and does not resolve with more time in bed. A child may fall asleep during meals, during a classroom lesson, or in the middle of a conversation. The sleep attacks feel irresistible and are not under voluntary control. The hallucinations, which occur at the edge of sleep onset or awakening, can be vivid and frightening, and children may have difficulty distinguishing them from real experience. Sleep paralysis, the temporary inability to move when waking, is frightening for both the child experiencing it and the parent watching it happen.

Pediatric narcolepsy typically emerges between the ages of seven and twenty-five, with a notable cluster of onset during puberty. The average time from symptom onset to diagnosis has historically been over a decade, which means many children spend years misunderstood before they get appropriate help. If your child shows a combination of severe, unexplained daytime sleepiness that does not resolve with adequate sleep time, plus any one of the other features described here, talk to your pediatrician about a referral to a pediatric sleep specialist. Diagnosis involves a sleep study followed by a daytime test called the Multiple Sleep Latency Test, which measures how quickly and in what patterns a child falls asleep across five naps. This specialized test, performed in a sleep laboratory, is the gold standard for diagnosing narcolepsy.

Treatment for pediatric narcolepsy involves a combination of behavioral strategies and medication. Scheduled naps can help manage daytime sleepiness, and consistent sleep timing is important. Medications used in adults, including sodium oxybate and various stimulants, are used in children with narcolepsy in modified forms and doses, always under the guidance of a specialist. If this picture resonates with what you are seeing in your child, the most important thing you can do is pursue evaluation. This is one condition where watchful waiting is not the right strategy.

Parasomnias: A Map of the Night

Parasomnias are a family of sleep disorders involving abnormal behaviors, movements, emotions, or experiences during sleep or during the transitions into and out of sleep. They sound dramatic, and they can certainly look dramatic. Still, most of them are developmentally normal in young children, are outgrown without intervention, and do not indicate psychological problems or stress unless they are unusually frequent, severe, or persist into adolescence.

Understanding where in the sleep architecture a parasomnia originates is key to understanding its character and how to respond.

Diagram 11.5: Parasomnias Map by Sleep Stage

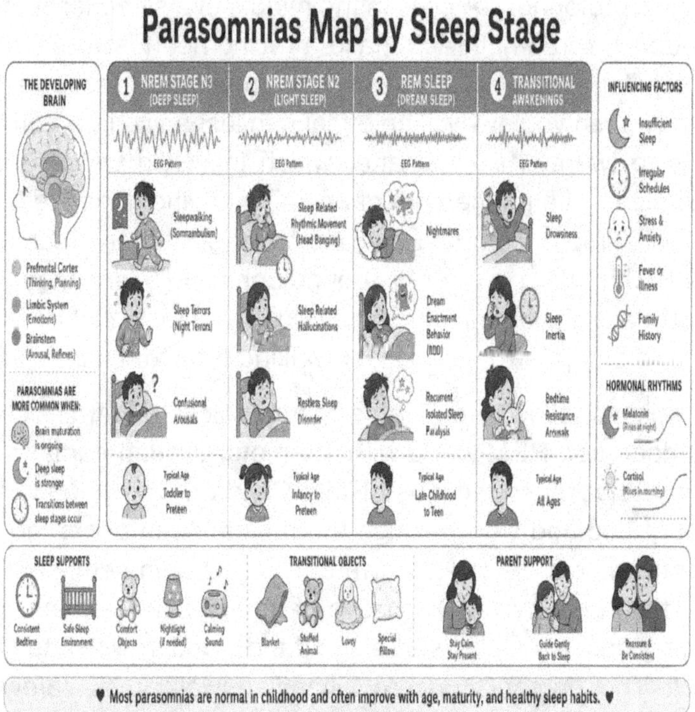

Non-REM parasomnias occur during deep slow-wave sleep, typically in the first third of the night. They include sleep terrors, sleepwalking, and confusional arousals. Because they arise from deep sleep, the child is genuinely not conscious during these events. They cannot be reasoned with, calmed by explanation, or guided by normal communication. Their eyes may be open, and they may even speak, but they are not awake, and they will not remember the event in the morning. This is one of the most counterintuitive things for parents to absorb: your child may look awake and be completely inaccessible. Trying to wake them fully or reason with them during the event typically prolongs it and increases the child's agitation.

Sleep terrors are the most alarming of the non-REM parasomnias, and they are worth understanding in detail because they frighten parents so profoundly. A child in the middle of a sleep terror may sit up suddenly, scream, thrash, have a racing heart, sweat, and look absolutely terrified. They may call out words but not respond to you. They may push you away. The event can last anywhere from a few minutes to twenty minutes, after which the child typically settles back to sleep with no memory of anything having happened. Sleep terrors are most common in preschool and early school-age children, with peak prevalence around ages 3 to 5, and they tend to diminish naturally by middle childhood. The family history is often significant; sleep terrors and sleepwalking tend to run in families.

Sleepwalking follows the same neurological pattern, occurring during deep non-REM sleep with the motor system partially active while consciousness remains offline. Children who sleepwalk may get out of bed and walk through the house, open doors, go to the bathroom without being fully aware, or engage in semi-purposeful but confused behaviors. Their eyes are typically open but glassy. They respond poorly to verbal guidance and resist attempts to interact. The primary concern with sleepwalking is safety. The environment should be made as safe as possible: stair gates in place, exterior doors locked or alarmed, sharp objects and hazards cleared from pathways, and the child's room arranged to minimize the risk of injury if the child gets out of bed. Gently guiding a sleepwalking child back to bed without fully waking them is usually the most effective approach.

Confusional arousals are similar in origin but typically less dramatic. The child partially wakes, appears confused and disoriented, may seem distressed, and is difficult to arouse fully. Unlike sleep terrors, confusional arousals are quieter and often shorter, though they can recur multiple times in a night. They are very common in infants and toddlers and represent an immature ability to navigate the transition from deep sleep to lighter stages.

Diagram 11.6: Bedwetting Decision Tree

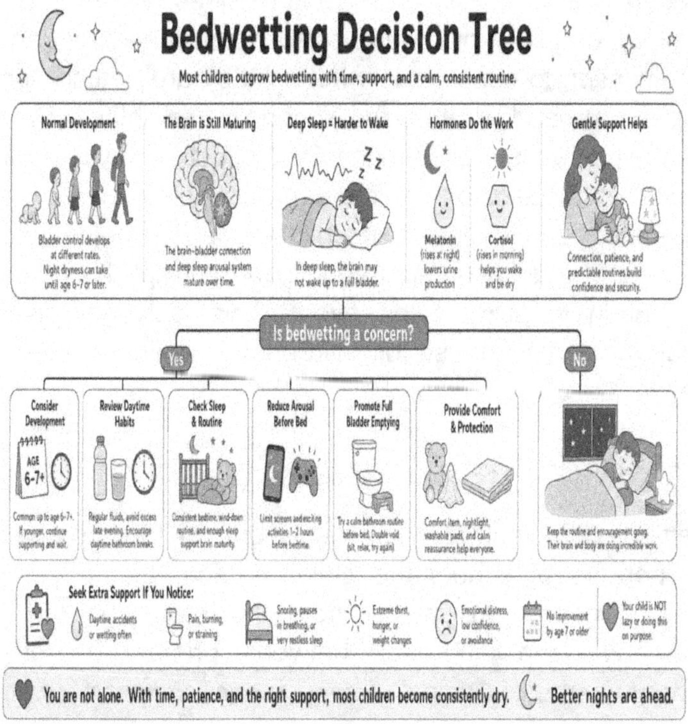

REM behavior disorder is a parasomnia that occurs during REM sleep rather than non-REM sleep, and it is worth mentioning specifically because it is rare in children and its presence in a child warrants medical evaluation. In REM sleep, the body is normally paralyzed through a mechanism called REM atonia. In REM behavior disorder, that atonia is absent or incomplete, and the person physically acts out their dreams. In adults, REM behavior disorder is associated with neurodegenerative diseases and is a cause for neurological evaluation. In children, it is sometimes seen in the context of narcolepsy, autoimmune conditions, or certain medications. If your child appears to be physically acting out vivid dreams, with purposeful-seeming movements, vocalizations, and apparent distress, talk to your pediatrician. This is not the same as a sleep terror, which occurs during non-REM sleep.

All non-REM parasomnias in children are worsened by sleep deprivation, fever, illness, and irregular sleep schedules. If your child's sleepwalking or night terrors are increasing in frequency, the first intervention to consider is whether they are getting adequate total sleep time and whether the schedule has shifted. Protecting overall sleep adequacy is often the most effective way to reduce the frequency of parasomnia episodes. Scheduled awakenings, in which a parent gently rouses the child about 30 minutes before the typical time of a parasomnia event, can interrupt the deep sleep cycle enough to prevent the event. This technique has good evidence supporting its use for children with frequent sleep terrors or sleepwalking.

Bedwetting: When to Wait, When to Act

Bedwetting, or nocturnal enuresis, is one of the most common issues parents of school-age children face, and it carries an outsized emotional weight for both the child and the family. Understanding the difference between primary enuresis and secondary enuresis helps clarify when watchful waiting is appropriate and when a pediatric visit is warranted.

Primary nocturnal enuresis refers to a child who has never achieved consistent nighttime dryness. It is the most common form and is largely a developmental issue involving delays in the maturation of bladder capacity, the nighttime arousal response, and the hormonal regulation of urine production during sleep. The vasopressin surge that normally suppresses urine production at night develops on its own schedule, and in some children that schedule runs later than average. About 15 percent of five-year-olds wet the bed. By age seven, the number drops to around 10 percent, and by age ten it falls further. Most children outgrow primary enuresis without any intervention beyond patience and protection. Waterproof mattress covers, low-key nighttime awakening routines, and a calm, shame-free family response are the foundations of management for most cases of primary enuresis.

Secondary nocturnal enuresis is different. It refers to a child who was previously dry at night for at least six months and has returned to bedwetting. Secondary enuresis is a flag. It can be triggered by a stressful life event such as a move, a new sibling, or a family disruption. Still, it can also signal a urinary tract infection, constipation, diabetes, or, as noted earlier, sleep apnea. Secondary enuresis warrants a visit to the pediatrician to rule out medical causes before assuming it is purely stress-related.

If a child is seven or older and still experiencing primary enuresis, it is also appropriate to bring it up with the pediatrician, both to assess for any contributing factors and to discuss the available interventions if the child is motivated to address it. Bedwetting alarms, which work by conditioning the child to wake in response to the first sensation of wetness, have the strongest evidence base of any behavioral intervention. They require consistency and patience but produce lasting results. Desmopressin, a synthetic form of vasopressin, is an effective short-term pharmacological option for situations where dryness is important, such as a sleepover or camp. However, it does not produce the same lasting effect as alarm therapy.

Bruxism: Teeth Grinding in Children

Bruxism, the habitual grinding or clenching of teeth during sleep, is surprisingly common in children, with some surveys finding it in 15 to 20 percent of school-age kids. It often occurs during lighter sleep stages or during transitions between sleep cycles, and in many children it is transient, appearing during periods of developmental change or stress and resolving on its own.

Parents typically first notice bruxism because the sound of grinding is audible across the room, or because a dentist notes wear patterns on the child's teeth at a routine appointment. In mild cases, no treatment is needed beyond monitoring and reassurance. In more significant cases, particularly when the grinding is causing dental damage or jaw pain, a custom-fitted night guard can protect the teeth. Dentists who work with children are experienced in this and can advise on timing and fit.

Persistent or worsening bruxism can be associated with stress, anxiety, airway obstruction, or RLS. If bruxism is accompanied by other sleep concerns such as snoring, restless sleep, or behavioral changes, it is worth mentioning all of the symptoms together at a pediatric appointment rather than treating the grinding in isolation. A dentist can protect the teeth, but identifying and addressing the underlying driver, whether it is anxiety, sleep apnea, or an iron deficiency, is where the real change happens.

Sleep-Related Anxiety vs. Sleep-Related Medical Issues

One of the harder diagnostic distinctions in pediatric sleep is telling the difference between anxiety that manifests at bedtime and sleep-related medical problems. They can look remarkably similar on the surface. A child with sleep apnea who wakes frequently throughout the night develops a secondary anxiety about sleep. A child with RLS develops fear about going to bed because they know how uncomfortable lying still will feel. A child with narcolepsy becomes anxious about having episodes in social settings. And children with entirely normal physiology can develop genuine anxiety disorders that make falling and staying asleep profoundly difficult. The overlap is real, and the lines are not always clean.

Diagram 11.7: When to See a Sleep Specialist

Some patterns help distinguish the two. Sleep-related anxiety typically follows a predictable situational pattern: the child is anxious specifically about sleep, about the dark, about being alone, about something bad happening. The anxiety activates at bedtime and usually settles once the fear trigger is removed, such as when a parent lies down with the child or a light is left on. The child typically falls asleep eventually and sleeps through without major disruption. Sleep quality, when it occurs, is generally adequate.

Medical sleep disorders tend not to follow that pattern. Obstructive sleep apnea disrupts sleep throughout the night, not just at the onset. RLS produces discomfort regardless of the child's emotional state and is relieved by movement rather than by reassurance. Narcolepsy produces daytime sleepiness regardless of how well the bedtime routine goes. These conditions do not respond to behavioral or emotional interventions because they are not driven by behavior or emotion. If your efforts to reduce anxiety and improve the sleep environment have not produced meaningful change, or if the daytime symptoms are significant regardless of how the night went, the remaining explanation is likely physiological rather than psychological.

That said, the two can absolutely coexist, and treating one does not guarantee resolution of the other. A child with OSA who has surgery may still have residual bedtime anxiety that developed during years of difficult sleep. A child with anxiety that was the original driver may have also developed secondary sleep fragmentation that takes time to resolve. Patience and a comprehensive approach that addresses both the medical and the emotional components serve these children best. Talk to your pediatrician about both dimensions so that nothing gets missed or attributed entirely to the wrong cause.

When to Consult a Pediatric Sleep Specialist

Most childhood sleep challenges can be navigated with the guidance of a knowledgeable pediatrician and the strategies in this book. But some situations are beyond the scope of primary care or parental intervention alone, and knowing when to escalate is important. Waiting too long to seek specialized evaluation is one of the most common missed opportunities in pediatric sleep medicine.

The following situations are clear indicators that a pediatric sleep specialist should be involved. Habitual snoring accompanied by witnessed apneas, gasping, labored breathing during sleep, or significant daytime behavioral or attention problems warrants a sleep evaluation. Excessive daytime sleepiness that is persistent, severe, and not explained by inadequate sleep time requires evaluation for narcolepsy or other hypersomnias. Cataplexy, or any episode that resembles a sudden loss of muscle tone triggered by emotion, should be evaluated promptly. Very frequent parasomnias, involve injury risk, persist into adolescence, or show any features of REM involvement rather than non-REM involvement warrant specialist input. Persistent insomnia in a child that has not responded to consistent behavioral intervention over several weeks, particularly when accompanied by daytime dysfunction, benefits from a specialized behavioral sleep medicine assessment.

Diagram 11.8: What a Pediatric Sleep Study Looks Like

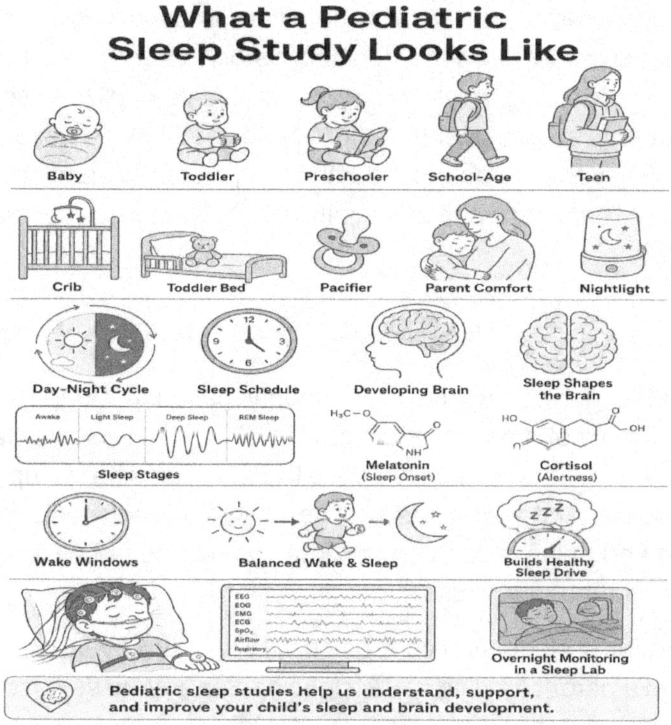

Restless legs symptoms with a strong family history, low ferritin, or significant bedtime distress that interferes with sleep deserve evaluation. Any child with an existing neurological condition, autism spectrum disorder, ADHD, or developmental delay who has significant sleep problems benefits from a sleep specialist consultation, because sleep-wake regulation is often more complex in these populations and requires tailored approaches. And any parent who has been told their child's sleep is fine but cannot shake the feeling that something is off deserves to be heard. Trust your instincts and advocate for the evaluation if you believe it is warranted.

A pediatric sleep specialist is typically a pediatrician with additional fellowship training in sleep medicine, though specialists may also come from neurology, pulmonology, or psychiatry backgrounds. Many academic children's hospitals have dedicated pediatric sleep centers. A referral from your pediatrician is usually the first step, and most specialists want a detailed history before recommending testing, so keeping a sleep log for a week or two before the appointment is genuinely useful. Note bedtimes, wake times, nighttime events, snoring, daytime symptoms, and anything unusual. That information significantly shortens the path to diagnosis.

What a Pediatric Sleep Study Involves

A pediatric polysomnogram, commonly called a sleep study, is nothing like what most parents imagine. It is not a hospital stay, it is not painful, and it is not as disruptive to the child as the setup makes it sound. Understanding the process helps parents prepare their children and helps reduce the anxiety that the idea of a sleep study can understandably create.

Most pediatric sleep studies are conducted in the evening, with the child and a parent arriving at the sleep center around seven or eight. The sleep center is typically a quiet room that looks more like a reasonably comfortable hotel room than a clinical space. Many centers have worked hard to make their pediatric rooms feel welcoming, with colorful decor, space for a parent to sleep nearby, and an effort to put children at ease during setup.

The setup involves attaching sensors to the child's scalp, face, chest, abdomen, and legs, as well as a pulse oximeter on a finger and sometimes a microphone near the throat. The sensors are held in place with medical paste or gentle adhesives and connected to lightweight wires that run to a recording box. The process of applying the sensors typically takes about thirty to sixty minutes and requires patience from both the child and the technician. For young children especially, the setup period can be the most challenging part of the night. Bringing a familiar stuffed animal, a favorite book, and familiar pajamas helps. Some families find it useful to talk to their child about the sensors as "stickers that help the doctor listen to your body while you sleep."

Once the child is asleep, the sensors record a remarkable amount of information simultaneously: brain activity via EEG, eye movements, muscle activity, heart rhythm, breathing effort and airflow, oxygen levels, leg movements, and body position. A technologist monitors the recordings throughout the night and may make small adjustments if a sensor comes loose. The child typically does not notice, and most children sleep reasonably well after an adjustment period. In the morning, the family is discharged, and the recordings are sent to the interpreting physician, typically a sleep medicine specialist, who manually scores the study against established criteria and produces a report.

The report takes time, often one to two weeks, and it documents the key metrics: the apnea-hypopnea index (AHI), oxygen saturation levels, sleep stage distribution, and any abnormal events recorded during the night. The referring physician or sleep specialist then discusses the findings with the family and outlines the recommended next steps. A single sleep study is usually sufficient for diagnosing OSA. Narcolepsy requires a two-day protocol that includes the nighttime polysomnogram followed immediately by the daytime Multiple Sleep Latency Test, because the daytime test must follow the sleep study directly to be valid.

The process is not as daunting as it sounds, and the information it provides is genuinely valuable. Parents often report that the hardest part was the anticipation, not the actual night.

Melatonin in Children: Use With Caution

Diagram 11.9: Melatonin: Use With Caution

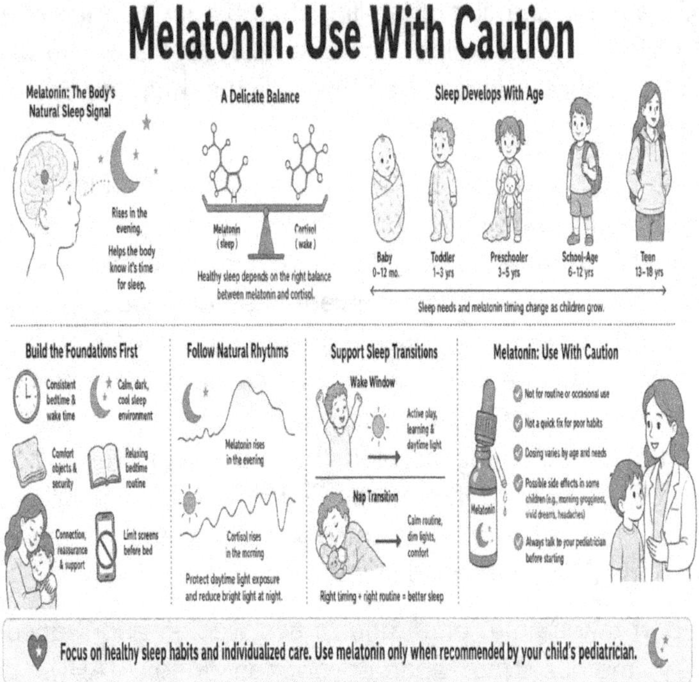

Pediatric melatonin use has increased dramatically over the past decade. Surveys suggest that millions of children in the United States are given melatonin regularly, often bought over the counter without guidance from a physician. The ease of access and the perception that it is a natural supplement have combined to make it the default first response for any child with sleep difficulties. The research, and the position of the American Academy of Pediatrics, suggest a more cautious and intentional approach.

Melatonin is a hormone, not a supplement in the conventional sense. The body produces it endogenously, and the supplement form does not simply top up a deficit the way a vitamin might. It signals to the circadian system that night has arrived, and its appropriate use in children is genuinely context-specific. The clearest evidence for melatonin's effectiveness in children is in two scenarios: circadian phase shifting in children with delayed sleep phase (where the circadian clock runs late and the child cannot fall asleep at a socially appropriate time), and sleep initiation in children with autism spectrum disorder or other neurodevelopmental conditions where endogenous melatonin production or timing is disrupted. In these contexts, low doses given at the appropriate circadian time can be meaningfully helpful.

For neurotypical children with behavioral sleep challenges related to routines, environment, anxiety, or parental limit-setting, melatonin does not address the underlying issue and its benefits are modest. More importantly, the long-term effects of exogenous melatonin on the developing hypothalamic-pituitary axis are not yet fully characterized. Melatonin receptors are distributed throughout the brain and body and play roles in development, immune function, and reproductive maturation that extend far beyond sleep. The precautionary principle applies here, particularly in very young children.

The AAP recommends that melatonin use in children be supervised by a pediatrician, used at the lowest effective dose (typically 0.5 to 1 milligram for most children, far below the doses in many commercial products), used for short courses rather than indefinitely, and not used as a substitute for addressing underlying behavioral or environmental sleep issues. If you are already using melatonin with your child, that is a conversation worth having at your next pediatric appointment. If you are considering starting it, talk to your pediatrician first rather than beginning on your own. The conversation takes ten minutes and ensures that the context, dosing, timing, and duration are appropriate for your specific child.

Prescription sleep aids are very rarely used in otherwise healthy children outside of specific clinical contexts. Antihistamines like diphenhydramine, which are present in many over-the-counter sleep products, produce sedation rather than physiological sleep and are not recommended for children as sleep aids. Clonidine and certain other medications are used in specific populations, such as children with ADHD, under physician supervision. None of these are appropriate for parental self-prescription. If you feel your child's sleep difficulties are significant enough to warrant medication, that determination belongs in a clinical conversation, not a pharmacy aisle.

The most durable solutions for pediatric sleep difficulties remain consistent timing, a supportive bedtime environment, responsive parenting that reduces anxiety around sleep, and, where medical conditions are present, appropriate diagnosis and treatment. Melatonin can be a helpful bridge in specific situations. It is not a fix for the full range of pediatric sleep challenges, and it works best when it is one part of a thoughtful, physician-guided plan.

Try This Tonight

For this age range: 5 to 18 years

1. Review the warning signs for pediatric OSA: habitual snoring on most nights, open-mouth breathing during sleep, witnessed pauses or gasping, and daytime behavioral symptoms like hyperactivity, inattention, or difficulty waking. If you recognize this pattern, write a pediatric appointment on your calendar tonight.

2. Think about whether your child's sleep difficulty follows the anxiety pattern (fear-driven, situational, resolves with reassurance) or the medical pattern (present regardless of emotional state, accompanied by physical symptoms, not responsive to behavioral change). That distinction helps you have a more productive conversation with your pediatrician.

3. If your child has had a parasomnia event, note the time it occurred relative to bedtime. Non-REM parasomnias almost always happen in the first third of the night during deep sleep. That timing, along with the description of the event, helps a physician make sense of what is happening.

4. If you are currently giving your child melatonin without physician guidance, make a note to bring it up at your next pediatric visit. Bring the bottle so the pediatrician can see the dose and formulation. A brief conversation can confirm whether what you are doing is appropriate or requires adjustments.

5. If your child is older than seven and still wetting the bed regularly, schedule a visit to the pediatrician specifically to discuss it. There is no shame in this conversation, and there are effective interventions when the timing is right and the child is motivated.

6. If you have a gut feeling that something is genuinely wrong with your child's sleep beyond behavioral issues, write down the specific observations you have made, with dates and descriptions where possible, and bring that list to your pediatrician. Concrete observations are more actionable than a general sense of concern, and your documentation is a real clinical tool.

Parent Takeaways

- Snoring on most nights, especially when it comes with daytime behavioral symptoms or unrefreshing sleep, is the most important single flag for pediatric OSA. This condition is common, treatable, and worth catching early.

- Pediatric sleep disorders frequently look like behavioral problems. Hyperactivity, inattention, emotional dysregulation, and defiance can all be driven by fragmented or disrupted sleep rather than by temperament or parenting.

• Parasomnias in young children are usually developmental and self-limiting. Safety-proof the environment, avoid waking the child during an event, and track frequency in case a medical evaluation becomes useful.

• Melatonin is a hormone with a specific and limited role in pediatric sleep. Talk to your pediatrician before using it, use the lowest effective dose for the shortest appropriate course, and do not let it substitute for addressing the underlying issue.

• Trusting your instincts matters. If you sense that your child's sleep problems are beyond behavioral causes, push for an evaluation. A pediatric sleep specialist exists precisely for this kind of question, and getting to the right diagnosis is always worth the effort.

12 Children's Sleep Myths Decoded

It is past midnight, and you are wide awake, not because your child is crying right now but because your mind is running a highlight reel of every piece of advice you have received in the past week. Your mother-in-law says you are spoiling the baby by going in too fast. Your best friend says you are damaging the bond by not going in fast enough. A post in a parenting group insists melatonin gummies are the only thing that saved their family. Another post, three comments later, says melatonin permanently disrupts the teenage brain. A pediatrician once told you bedsharing was dangerous. Someone else told you bedsharing is how the rest of the world does it and American medicine is fear-mongering. A parenting book you started in your third trimester told you that a good baby sleeps through the night by six weeks. Your baby is twelve weeks old and wakes up four times, and you have started to wonder privately whether there is something wrong with her.

There is nothing wrong with her. There is something wrong with the advice.

Here is the thing about children's sleep myths: they do not feel like myths. They feel like wisdom. They arrive wrapped in the authority of a grandparent's experience, the social proof of a friend's apparent success, or the calm confidence of a stranger online who seems very sure of themselves. The myths survive because they are passed from adult to adult in the same breath as genuine care, and because the real science is harder to find than a Facebook group, quieter than a confident neighbor, and far less satisfying than a simple rule.

This whole book has been building toward this chapter. Every concept you have read, every diagram you have examined, every "Try This Tonight" you have attempted, has been giving you the background to hear the myths and recognize what is actually being claimed. You are no longer starting from zero. You understand sleep cycles, circadian biology, sleep pressure, the two-brain problem that every tired parent eventually faces when their own need for sleep collides with their child's needs at two in the morning. That understanding is your armor.

This final chapter walks you through twelve of the most persistent, most harmful, and most widely believed myths about children's sleep. Some of them are myths that cause real harm by making parents comfortable with a problem that deserves attention. Some cause harm by making parents feel guilty about doing something that is perfectly safe and evidence-supported. Some are myths of omission, partial truths that become dangerous when they are stripped of their context. For each one, we will look at what the science actually says, and why it says it.

You will probably recognize at least half of these from your own experience. You may have believed some of them. That is not a character flaw. That is what happens when culture moves faster than science and when everyone around you is exhausted enough to repeat the last thing that felt like an answer. The goal of this chapter is not to make you feel foolish for having believed something. The goal is to give you the clearest possible picture of what is actually true, so that the next myth you encounter, the one that has not been invented yet, does not land unopposed.

Chapter Summary

This capstone chapter takes the twelve most common and most consequential myths about children's sleep and dismantles each one against the developmental neuroscience this book has built. It covers myths that cause harm by generating unwarranted guilt, myths that cause harm by offering false reassurance, and myths that fall somewhere in between, where a partial truth is real but the stripped-down version people actually share is misleading.

The chapter closes by returning to the Two-Brain Problem introduced in Chapter 1, because that concept turns out to be the thread running beneath almost every myth on this list. When you understand that your brain and your child's brain have genuinely different biological needs and genuinely different developmental timelines, most of the myths collapse on their own. You stop asking whether you are doing it right and start asking what your particular child needs tonight, and whether you are in good enough shape to provide it.

What This Chapter Will Cover

- Myth: Good babies sleep through the night early. The biology of infant sleep architecture explains why this expectation is not just wrong but potentially counterproductive, as it can drive parents to override genuine hunger or comfort needs in very young infants.
- Myth: You should never let them cry. This chapter presents the evidence on extinction and graduated extinction methods honestly, without the guilt-loading that usually surrounds this topic, and without the evangelical enthusiasm that sometimes appears on the other side.
- Myth: Crying it out causes lasting trauma. Multiple well-designed long-term studies have followed infants through sleep training into childhood. This chapter summarizes what those studies actually found, which is considerably less alarming than what gets shared on social media.
- Myth: Bedsharing is always dangerous. The American Academy of Pediatrics guidance exists for important reasons. So does the anthropological reality that most humans throughout history and most cultures today do not sleep alone. This chapter explains the risk spectrum and the conditions that matter most.

• Myth: Skipping naps means better night sleep. This is one of the most intuitive-feeling myths on the list and one of the most reliably wrong. The neuroscience of sleep pressure and cortisol explains why the opposite tends to be true.

• Myth: TV helps kids wind down at bedtime. The argument sounds logical: passive, relaxed content, child lying still. The reality is that screen exposure at bedtime is working against sleep biology on at least three separate biological channels simultaneously.

• Myth: Teens are just lazy sleepers. Adolescent sleep phase delay is a well-documented biological shift driven by changes in both circadian timing and sleep pressure. It is not a character trait. It is not a discipline problem. It is puberty, expressed in the circadian system.

• Myth: Melatonin is harmless for kids. Melatonin has become one of the most widely used pediatric supplements in the United States. The short-term picture is reasonably reassuring in specific contexts. The long-term picture is largely unknown, which is itself a reason for caution.

• Myth: Sleep training breaks the parent-child bond. This is the myth that causes the most unnecessary guilt, because it targets something parents care about more than almost anything else. The evidence from attachment research is clear and worth knowing in full.

• Myth: My kid will grow out of it. Sometimes true. Sometimes the symptom being dismissed is obstructive sleep apnea, a treatable condition with measurable cognitive consequences when left unaddressed.

• Myth: More sleep is always better. Sleep needs sit within a healthy range, not a one-directional maximum. Both too little and too much can signal or create problems, and chasing more sleep past the upper bound is not a strategy.

- Myth: The perfect schedule exists. It does not. What exists is a flexible framework that you adjust continuously as your child grows, as life changes, and as you learn what your particular child's nervous system actually needs.

Diagram 12.1: The Myth of Sleeping Through the Night

Myth One: Good Babies Sleep Through the Night

You hear some version of this almost every week in the first year. It usually comes packaged with a comparison: "My sister's baby was sleeping eight hours straight by six weeks." It lands like a verdict on your child, on your parenting, on something.

The biology is clear, and it does not support the verdict. A newborn's sleep cycle runs about 50 minutes, compared to the 90-minute adult cycle. Newborns spend roughly half their sleep in active sleep, the developmental predecessor to REM, and arousals at the end of each cycle are not a malfunction. They are a feature. The arousal threshold in early infancy is protective. Newborns have tiny stomachs, high caloric demand relative to body weight, and developing nervous systems that are not yet capable of consolidating sleep the way adult systems do. Expecting a six-week-old to sleep through the night is, biologically, like expecting a newborn to walk.

Most infants develop the neurological capacity for longer, more consolidated stretches between four and six months, though the range is wide. By 12 months, a majority of healthy infants can sleep longer stretches, but even then, "sleeping through the night" in research terms usually means 5 or 6 consecutive hours, not the 8 or 9 hours adults tend to imagine when they use that phrase. Night waking persists in a meaningful percentage of toddlers and even school-age children, and in most of those cases it reflects normal variation in nervous system development, not a problem that needs to be solved by twelve weeks.

The harm in this myth lies not just in creating false expectations. The harm is that it makes parents of normal infants feel like something has gone wrong, which increases parental anxiety, which disrupts parental sleep in an independent and additional way, which affects the quality of daytime caregiving. You worry; your body shows it, and the baby reads it. Releasing this myth is not lowering the bar. It is removing a bar that should never have been there.

Diagram 12.2: The Evidence on Extinction Methods

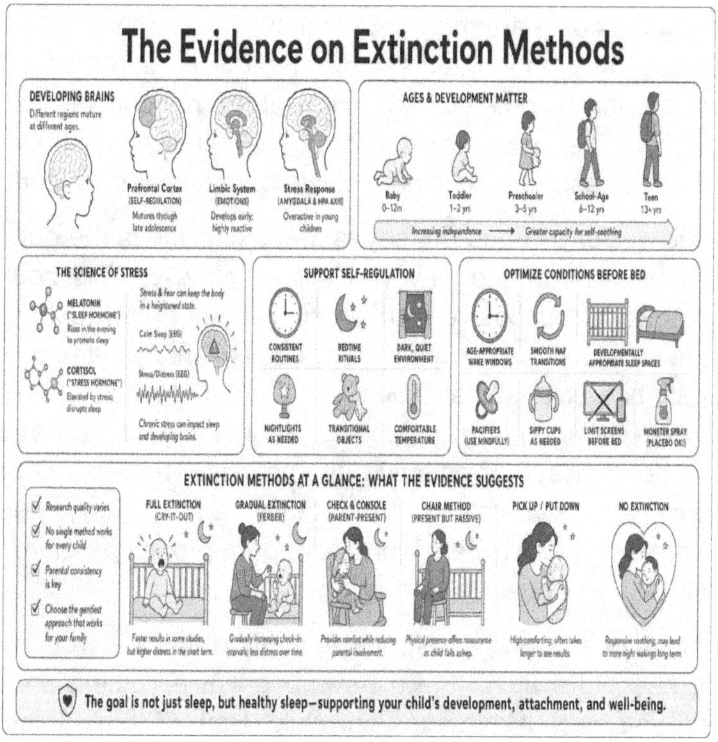

Myth Two: You Should Never Let Them Cry

This one arrives with tremendous emotional force because it weaponizes your instincts. Your instinct to respond to a crying baby is not a bug. It is a feature of primate parenting that is millions of years old and deeply adaptive. The question is not whether that instinct is valid. It is whether the extrapolated rule that any unanswered cry is evidence of harm supports it.

The evidence does not support it. That is not a political statement. It is a finding.

Extinction methods, which include classic cry-it-out approaches as well as graduated-extinction approaches like the Ferber method, have been studied in multiple randomized controlled trials and followed up over months and years. The consistent findings are that these methods reduce night waking, that parents who use them report reduced parental depression and improved family functioning, and that infants who underwent behavioral sleep training are indistinguishable at follow-up from those who did not. Indistinguishable on stress hormones. Indistinguishable on behavioral measures. Indistinguishable on attachment security.

What the evidence does not say is that extinction is the only way, or the best way, or the way you should choose. This book has leaned toward responsive and gradual methods throughout, not because extinction is harmful, but because many parents find it genuinely harder to sustain, because shorter-term distress is real even when the long-term outcomes are equivalent, and because there is no shortage of effective alternatives. The best method is the one you can follow consistently and calmly, because inconsistency in any behavioral sleep approach is harder on children than any specific method.

If you have been using a responsive method and it is working, you do not need to switch. If you have been using extinction and it worked, you did not damage your child. Both of those things are true at the same time.

Diagram 12.3: Bedsharing Risk vs Reality Spectrum

Myth Three: Crying It Out Causes Lasting Trauma

This myth is worth its own section because it is the mechanism behind a great deal of parental guilt, and because it is stated with such confidence in some parenting communities that parents who have used behavioral sleep training sometimes carry genuine distress about it for years.

The longitudinal evidence does not support the trauma claim. The most rigorous studies, including a randomized trial that followed infants for 5 years after behavioral sleep training, found no differences in emotional and behavioral outcomes, attachment security, stress reactivity, or parent-child relationship quality. The babies who cried for some nights at twelve months became five-year-olds who were statistically identical to the babies who did not.

That does not mean the nights were comfortable for anyone. Crying is distress in the moment, and parents who sat outside a door listening were experiencing their own real distress. The evidence indicates that the acute experience did not result in chronic developmental harm. It does not say it felt easy or neutral.

The trauma claim, when examined carefully, often rests on a conflation of two different things: the stress response that occurs during the extinction process, and the chronic, unpredictable, caregiver-unresponsive stress that produces the kind of lasting neurological changes associated with adverse childhood experiences. Those are not the same thing. Brief, bounded periods of distress in the context of a warm, responsive caregiving relationship do not function neurobiologically the way chronic, unpredictable neglect does. The distinction matters enormously.

If you tried behavioral sleep training and feel conflicted about it, what the evidence actually offers you is reassurance, not continued uncertainty. Your child is fine.

Myth Four: Bedsharing Is Always Dangerous

This myth lives at one end of a genuinely complicated conversation. The American Academy of Pediatrics recommends against bedsharing for infants, particularly for the first six months of life, and particularly under conditions that increase SIDS risk. That guidance reflects real data and serves real purposes. The risk is not invented.

The risk is also not flat. It varies dramatically by context, and understanding the variables is more useful than a blanket rule because many families around the world bedshare, and many families in the United States do so despite recommendations. Families who are going to bedshare are better served by clear information about the risk factors than by a one-size-fits-all warning they may already have chosen to disregard.

The factors that most significantly elevate bedsharing risk include soft bedding, parental smoking (even if smoking does not happen in the bedroom), parental alcohol or sedating medication use, and extremely soft sleep surfaces. Remove those factors and the risk picture changes substantially, though it does not reach zero. The AAP guidance recommends room-sharing without bedsharing as a middle path that captures the protective effects of proximity, which are real and documented, without the bedsharing risk profile.

Room-sharing is associated with a reduced risk of SIDS. Proximity helps a breastfeeding parent respond quickly and may support breastfeeding duration, which is itself protective. Having the infant in your room in a separate safe sleep surface gives you most of what proximity offers.

Diagram 12.4: Why Skipped Naps Hurt Night Sleep

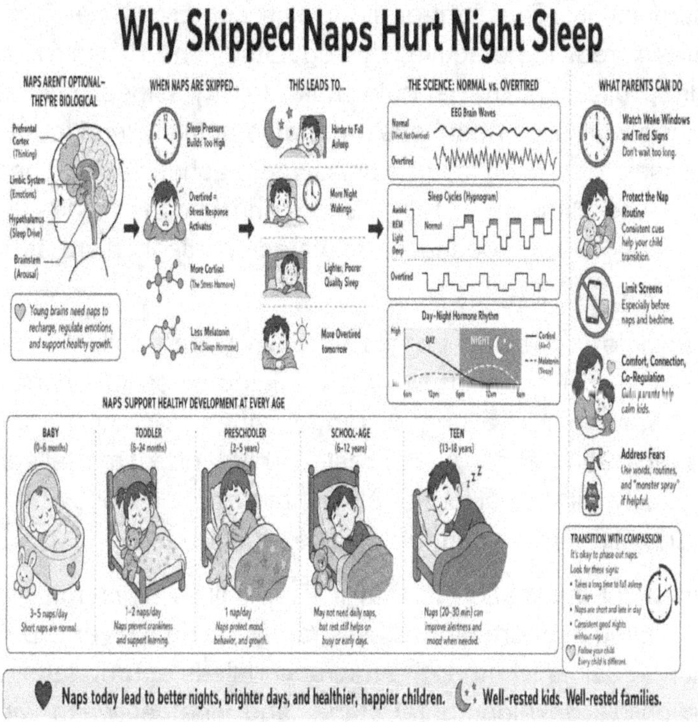

Myth Five: Skipping Naps Means Better Night Sleep

This one runs so counter to intuition that it deserves a slow walk. The logic behind it seems airtight: a child who skips a nap is more tired; a more tired child falls asleep faster; therefore, skipping the nap improves the night. Every step of that logic sounds right. Almost every step of it is wrong.

The problem is what happens between the skipped nap and bedtime. An overtired child is not simply tired. An overtired child is a child whose body has activated a cortisol and adrenaline response to compensate for the fatigue, because the brain interprets extended wakefulness as a demand state and responds with stimulation. The fight-or-flight axis comes online. The child becomes wired, hyperactive, dysregulated, and difficult to settle. The sleep pressure that should make falling asleep easy is now competing with an arousal system that is actively working against sleep onset. The result is not a child who crashes easily. The result is a child who is miserable to get to bed and who then tends to sleep worse, not better, because elevated cortisol disrupts sleep architecture.

The principle that sleep begets sleep in young children is not a slogan. It is a description of how pediatric sleep pressure works when it stays within a normal range rather than tipping into the cortisol-activation zone. A well-rested child at bedtime falls asleep more easily and maintains sleep more consistently than an overtired child. This is counterintuitive, and it is true.

The practical implication is that if your child's nighttime sleep is poor, skipping or shortening naps is rarely the right lever to pull. The question to ask is whether the nap schedule is appropriately timed for the child's developmental stage, and whether wake windows before the final nap and before bedtime fall within the zone that allows gentle, natural drowsiness without tipping into the cortisol response.

Diagram 12.5: Blue Light and Bedtime Arousal

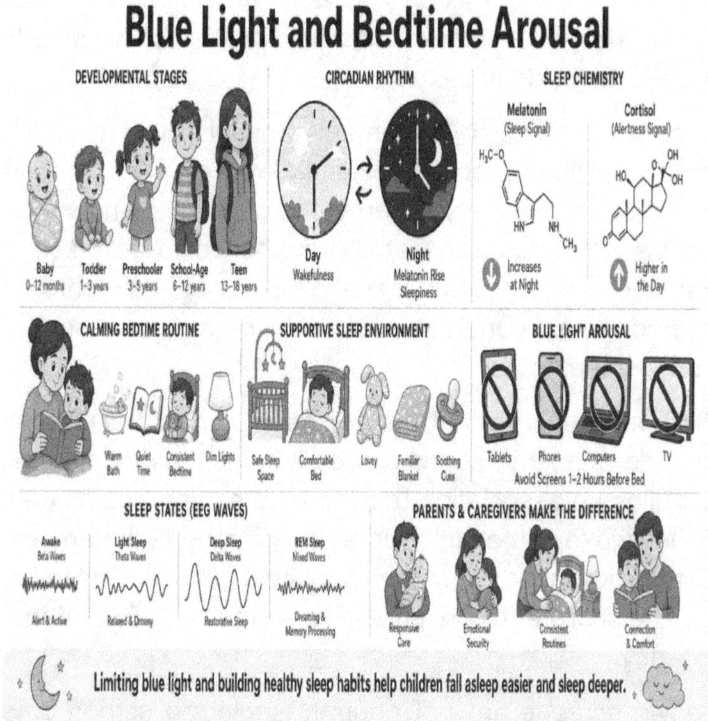

Myth Six: TV Helps Kids Wind Down

The mental image behind this myth is a child lying passively on a couch, watching something calm, eyelids gradually getting heavy. It is a plausible image. It is not what is actually happening in the brain.

There are at least three distinct biological mechanisms by which screen exposure in the hour before bed actively disrupts sleep, and they operate simultaneously.

The first is blue-light suppression of melatonin. The screens in your home, including televisions, tablets, and phones, emit light in the wavelength range that the suprachiasmatic nucleus uses to calibrate circadian timing. Blue-spectrum light tells the brain it is daytime. It suppresses the melatonin rise that needs to begin roughly two hours before sleep onset. The effect is not metaphorical. In controlled studies, evening screen exposure measurably delays melatonin onset and pushes back sleep timing, even when the child feels tired.

The second mechanism is content arousal. Even calm content requires the brain to process social and narrative information, and it does not do so passively. Characters, storylines, and social cues engage the same circuits that are supposed to be quieting. For children who are particularly sensitive to emotional content, even gentle stories can generate sufficient neural activation to delay sleep onset meaningfully.

The third mechanism is habit association. A child who falls asleep with the television on or uses screen time as the primary transition to bed builds an association between the screen and sleep onset. When they wake in the night, as all children do at the end of sleep cycles, they now need the screen condition to be present to fall back asleep. The short-term convenience becomes a middle-of-the-night dependency.

Wind-down screens are not neutral. Replacing screen time with lower-stimulation activities, reading together, quiet play, a bath, dimmed lights, and consistent sensory cues is not arbitrary restriction. It is working with the biology that governs sleep onset.

Myth Seven: Teens Are Just Lazy Sleepers

This one does cultural harm at a scale that is hard to overstate, because it pathologizes a biological process as a character flaw, and it does so in front of teenagers who are already vulnerable to shame around their identities.

Adolescent sleep phase delay is one of the most well-documented and consistently replicated findings in sleep science. It is not laziness. It is a genuine shift in the circadian clock that occurs during puberty, driven by changes in melatonin timing and sleep pressure dynamics, and it is observed across cultures and species.

Adolescent rodents, adolescent primates, and adolescent humans all show the same shift. They naturally fall asleep later and naturally wake later than either children or adults.

The magnitude of the shift is roughly two hours. A teenager whose circadian system wants them asleep at midnight and awake at eight in the morning is not running on a preference or a habit. They are running on a biological clock that is doing exactly what biology instructs it to do during this developmental window.

School start times that require fifteen-year-olds to wake at six in the morning are, biologically, asking them to wake at what is effectively four in the morning on an adult clock. The chronic sleep restriction that results has measurable consequences: impaired memory consolidation, reduced emotional regulation, higher rates of depression and anxiety, worse academic outcomes, and higher rates of drowsy-driving accidents in older teens who drive. These are not soft concerns.

Diagram 12.6: The Teen Phase Delay Revisited

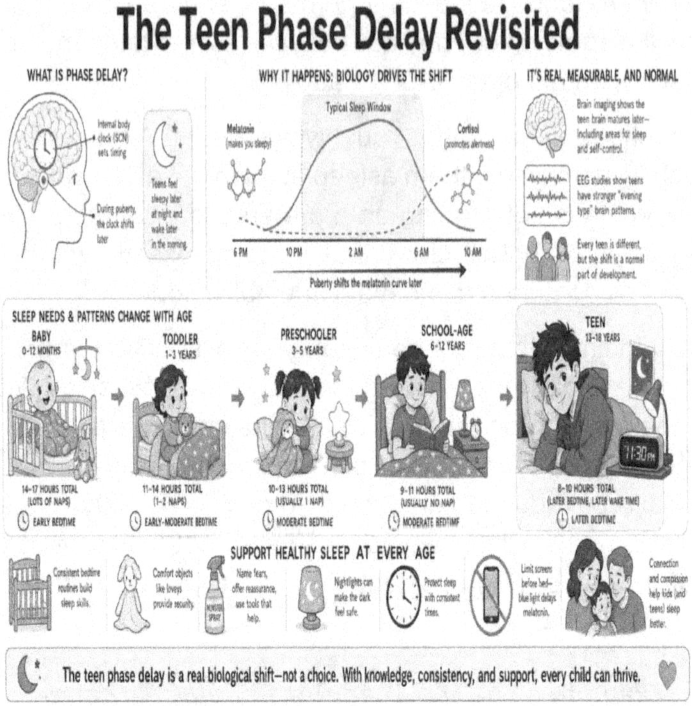

The response to the teenager who cannot fall asleep at nine and cannot wake up at six is not earlier bedtimes and earlier wake times, at least not initially. The response is a gradual, consistent, darkness-based realignment of the circadian clock over weeks, combined with advocacy for school schedules that reflect the biology, and a shift in cultural framing from "lazy" to "out of phase."

Teenagers who have been labeled lazy sleepers for years often discover, when they first learn about the phase delay, that they feel both relieved and angry. Relieved because it was not a character flaw. Angry because everyone around them behaved as though it was. That relief and that anger are both completely reasonable.

Myth Eight: Melatonin Is Harmless for Kids

Pediatric melatonin use has grown dramatically over the past decade. In the United States, melatonin is sold over the counter without age restrictions, marketed in child-friendly gummy forms, and widely recommended in parenting communities as a gentle and natural sleep solution. Most parents who use it assume it is safe because it is available without a prescription and because it is a hormone the body makes anyway.

The short-term picture is nuanced rather than alarming. For specific populations, particularly children with autism spectrum disorder or ADHD who have genuine circadian disruption, low-dose melatonin used under pediatric guidance appears helpful, and the short-term safety profile is reasonable. For typically developing children with behavioral sleep problems rather than circadian disruption, the American Academy of Pediatrics recommends addressing behavioral and environmental factors first, using melatonin as a last resort rather than a first-line option, and keeping doses as low as possible.

The long-term picture is largely unknown. Melatonin is involved in reproductive hormone signaling during puberty. Animal studies have shown effects on pubertal timing with long-term melatonin use. Human studies on long-term pediatric use are limited because the widespread consumer use is relatively recent. The absence of documented harm is not the same thing as established safety, and the honest answer for a parent asking about long-term melatonin use in young children is that we do not know yet.

Diagram 12.7: Melatonin Short-Term vs Long-Term

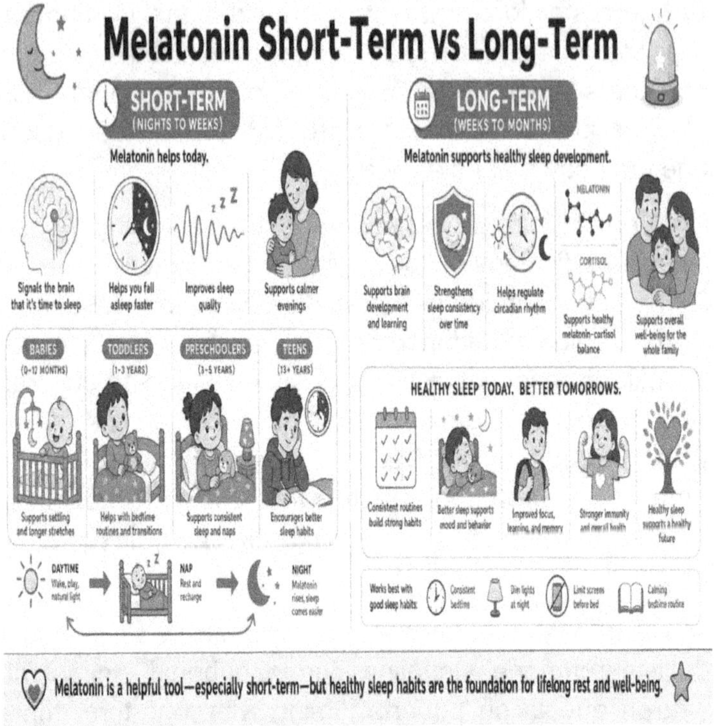

This is not a call to panic about melatonin use you have already done. It is a call to use the lowest effective dose, to use it for short courses rather than indefinitely, to involve your pediatrician in the decision, and to prioritize the environmental and behavioral levers- light management, consistent bedtimes, wind-down routines- first. Melatonin is a tool, not a substitute for a sleep environment.

Myth Nine: Sleep Training Breaks the Parent-Child Bond

Of all the myths in this chapter, this one generates the most email to sleep researchers. It generates that email because parents who are told this believe it deeply, act on it (either by avoiding sleep training and suffering longer than necessary, or by doing sleep training and carrying guilt about it afterward), and are often difficult to reassure because the claim lands in the attachment center of parenting identity, not in the part of the brain that weighs evidence.

The attachment research is detailed. Attachment security, measured through the gold-standard Strange Situation assessment, is determined by the overall pattern of caregiver responsiveness across thousands of interactions over the first years of life. It is not determined by a handful of nights of behavioral sleep training at eight months. The research includes randomized trials that measured attachment both before and after sleep training. The attachment scores do not change.

The intuitive argument against this finding usually goes as follows: if I leave my baby crying, I am teaching them that I will not come, which undermines their sense of security. That argument is coherent as far as it goes. What it misses is that attachment security is not calculated by the child from a single episode or a short series of episodes. It is built from a cumulative history of caregiving. A baby who has experienced a warm, responsive parent hundreds of times over the course of months does not rewrite that history because of several nights of sleep training. The secure base remains.

What behavioral sleep training does change is the child's expectation about nighttime specifically. That is a real change. But the evidence is consistent. This change in nighttime expectations does not generalize to daytime attachment, does not affect emotional development, and does not register in the measures that attachment science uses to assess the parent-child bond.

Diagram 12.8: Sleep Training and Attachment Security

Myth Ten: My Kid Will Grow Out of It

This one is sometimes true. And sometimes it is how obstructive sleep apnea in children goes undiagnosed for years while measurable cognitive and behavioral damage accumulates.

Normal developmental changes do resolve some sleep problems on their own. The four-month sleep regression passes. Many sleep associations fade as children mature. Night waking decreases in frequency across the toddler years for most children even without structured intervention. There is a reasonable watchful-waiting argument for many sleep difficulties in otherwise healthy children.

The symptom that should not be waited on is habitual snoring. Snoring in a child is not normal, unlike snoring in an adult, which has been culturally normalized. It is a sign of elevated airway resistance during sleep. In a meaningful percentage of children who snore habitually, that elevation meets criteria for obstructive sleep apnea, a condition in which the airway partially collapses repeatedly during sleep, fragmenting sleep architecture without the child or parent necessarily being aware of the mechanism.

Pediatric OSA is associated with impaired memory consolidation, behavioral dysregulation, attention problems that are frequently misdiagnosed as ADHD, and growth concerns. It is treatable, most often with adenotonsillectomy, or, in some cases, with other interventions. The outcomes after treatment are often dramatic: children who were struggling in school, described as difficult or hyperactive, and sleeping fitfully, become markedly different within months of having the airway obstruction resolved.

The rule worth holding is this: wait and see is reasonable for garden-variety sleep development. Habitual snoring, witnessed apnea (pauses in breathing), and significant daytime behavioral or cognitive concerns warrant a conversation with a pediatrician who should consider a referral for sleep evaluation. Your child will not necessarily outgrow an airway problem. They will grow older with it.

Myth Eleven: More Sleep Is Always Better

This myth is understandable given how relentlessly underslept most families are. If some sleep is good, more sleep is better, and the limit is higher than anyone ever reaches. It feels like a safe direction to push.

Sleep needs in children, like those in adults, fall within a healthy range rather than a single, fixed maximum. The ranges published by major pediatric and sleep organizations are based on outcome data on which levels of sleep duration are associated with healthy cognitive development, healthy emotional regulation, healthy immune function, and healthy growth. Consistently sleeping below the lower end of the range has been documented to have costs. Consistently sleeping above the upper end of the range is associated with some negative outcomes as well, including, in some studies, higher rates of depression and lower daytime functioning.

For most families, the relevant direction is still "more" rather than "less." True chronic long-sleeping in children is relatively uncommon and usually has a medical explanation worth evaluating, whether that is illness, medication effects, or a sleep disorder causing poor-quality sleep and the body compensating with increased quantity. The practical implication is that chasing sleep beyond the appropriate upper bound for your child's age is not a productive goal. The goal is sleep that is appropriate in quantity and good in quality, not sleep that is maximized without a ceiling.

Diagram 12.9: The Two-Brain Problem Revisited

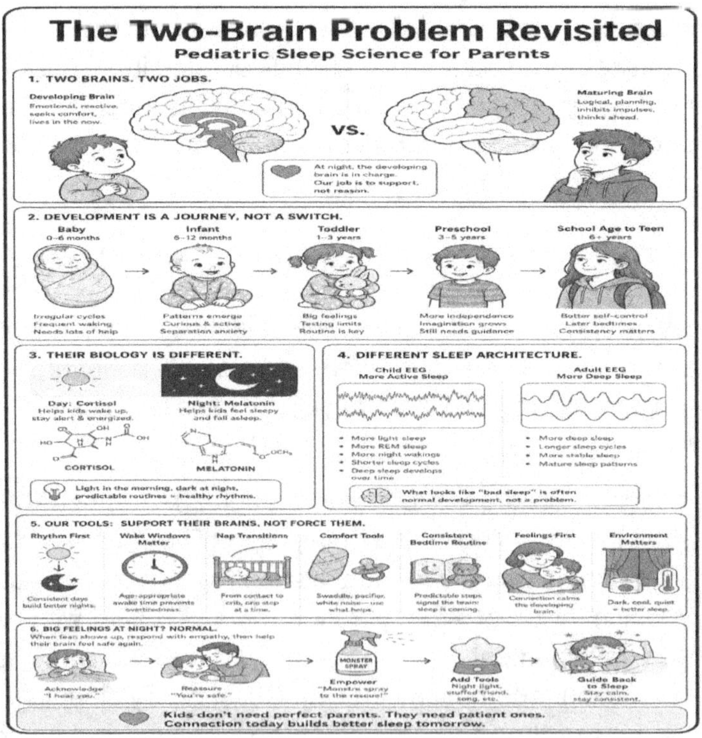

Myth Twelve: The Perfect Schedule Exists

Every parenting category has its version of the perfect schedule. There are apps. There are books. There are accounts with charts, wake windows, precise feeding intervals, and optimized nap durations. The schedules are presented with authority and followed by parents with something close to devotion. When the schedule does not work, parents almost universally conclude that they are implementing it wrong rather than that the schedule is wrong for their child.

The perfect schedule does not exist. What exists is a set of biological principles, developmentally appropriate wake windows, the role of light and darkness in circadian anchoring, the relationship between nap timing and nighttime sleep pressure, the cortisol-overtiredness curve, all of which you have now encountered across this entire book. These principles are real. They apply to every child. But the specific schedule that emerges from applying them to your particular child, in your particular home, with your particular work situation, feeding approach, and family structure, is yours to build. It is not downloadable.

Children vary in sleep need within age-normal ranges. They vary in sensitivity to overtiredness. They vary in how quickly their circadian systems respond to environmental cues. They go through developmental leaps, illnesses, travel, time changes, and family disruptions that require the schedule to flex. The parent who is rigidly attached to a particular schedule will experience every deviation as a failure, which is both exhausting and inaccurate. The parent who understands the underlying biology can adapt because they know why the schedule is structured as it is and which variables matter most when something needs to change.

Flexibility is not abandonment. Adjusting the schedule when your child is sick is not destroying the sleep training. Moving bedtime thirty minutes for a family gathering is not resetting the clock to zero. The goal was never a schedule. The goal was a child who has enough consistent, biologically appropriate sleep to grow well, think clearly, and feel like themselves, and a parent who has enough rest to be present for that child. The schedule is a vehicle, not a destination.

Try This Tonight

For this age range: All ages

1. Pick one myth from this chapter that has shaped how you think about your child's sleep, whether it created guilt, false reassurance, or confusion, and write it down in a single sentence.

2. Write one sentence describing what the science actually shows instead. You do not need to memorize the research. You need the corrected sentence to exist somewhere outside your head.

3. If you have a co-parent or another adult in the household, share both sentences. Aligned caregivers are the most powerful sleep intervention in any home, and this is one of the lowest-effort alignment tools available.

4. Identify whether any myth on this list has caused you to avoid a conversation with your child's pediatrician. If yes, write down the symptom or concern you have been dismissing. Habitual snoring belongs in that conversation. Dramatic behavioral dysregulation that resolves on weekends when sleep is longer belongs there too.

5. Look at your current bedtime routine and identify one element that came from advice rather than observation of your specific child. Spend one week running that element as an experiment rather than a rule, and notice whether removing or changing it makes any difference.

6. If you have a teenager in the house, share what you learned about the adolescent phase delay with them directly. Not as a

lecture. As information that belongs to them about their own biology.

7. Give yourself credit for the days and nights you showed up even when you were running on almost nothing. The evidence on what children need most is consistent: a present, regulated, warm caregiver matters more than any specific schedule, method, or technique. You have been that caregiver through imperfect nights, contradictory advice, and genuine exhaustion. That counts enormously.

Parent Takeaways

- Night waking in the first four to six months is biologically normal, not a sign of a problem with your child or your parenting.

- Sleep training methods, including extinction approaches, do not damage attachment security or cause lasting emotional harm when used in the context of a warm, responsive caregiving relationship.

- Habitual snoring in a child is not normal and warrants pediatric evaluation, because pediatric OSA is treatable and the costs of leaving it untreated are real.

- The adolescent phase delay is a biological event, not a character trait. Treating it as laziness generates shame without producing earlier sleep.

- Flexibility within a biologically grounded framework serves children and parents better than rigid adherence to any single schedule or method.

- The goal was never perfection. The goal was enough good sleep, consistently enough, for both of you to keep going.

This Is Where You Are Now

You started this book, or perhaps you started it weeks ago in a sleep-deprived fog with a newborn on your chest, looking for something solid to hold onto. What this book has tried to give you is not a set of rules but a set of tools. The rules change as children grow. The biology behind them does not.

The Two-Brain Problem from Chapter 1 is worth naming one more time, here at the end, because it is the context inside which everything else sits. You and your child are two separate nervous systems trying to coexist in a shared household where your sleep needs and theirs do not naturally align, at least not in the early years. Your child's sleep architecture is developing in real time. Their circadian system is immature. Their drive to stay connected to you is not a manipulation. It is a survival instinct that has served our species for a very long time and is wired deeply into both of your brains.

Your needs are also real. Your sleep deprivation is not a small thing. Your functioning, your mood, your physical health, your capacity to be the warm and regulated presence your child needs, all of it degrades under chronic sleep restriction. The parent who tends to their own sleep is not selfish. They are doing maintenance on the most important tool in the house.

The myths on this list survive partly because they give people simple answers to a genuinely complicated situation, and partly because exhausted people cannot always do the research, and partly because the stakes feel enormous and the desire for certainty is therefore enormous too. Now that you have the framework, you do not need certainty. You need enough understanding to make reasonable decisions for your specific child on specific nights, to know when something is developmentally normal and when it warrants attention, and to stop carrying guilt about things the evidence has long since cleared.

Your child does not need you to be perfect at this. They need you to keep showing up, to keep learning, and to know when to ask for help. You have been doing all three. That is what good parenting looks like when sleep is hard, the advice is loud, and the nights are long.

The nights do get longer in hours of sleep. They always do. You are closer to that than it probably feels right now.

13 A Final Note

There is a particular kind of trust that happens at three in the morning.

A parent picks up a book, or a phone, or a dog-eared printout from a pediatrician's waiting room. They are not reading for pleasure. They are reading because the child down the hall has been awake again, or because they finally got the child to sleep and now cannot sleep themselves, or because the question that has been sitting in their chest all day finally has enough quiet around it to surface. They are reading with tired eyes and a mind that keeps getting pulled in two directions at once: toward whatever is on the page and toward the small, breathing, unpredictable person sleeping a few feet away.

That trust is nothing. When you let a book into that moment, into the specific vulnerability of parental exhaustion at the tail end of a hard night, you are extending something genuine. The author on the other end of that exchange has a responsibility to meet it honestly.

I wrote this book inside that responsibility. My own exhaustion feels like a long time ago now, but it was real, and the particular quality of it, the way it lived not just in the body but in the self, the low-level doubt that shadowed every decision, is not something you forget. When I began the research that eventually became this series, I was not looking for the perfect method. I was looking for a framework that would let a tired parent understand what was actually happening, and then make their own call. I was looking for a book I would have wanted to read at three in the morning, one that did not require me to pick a team, did not assume I had failed, and did not treat my own sleep as a luxury detail.

This is that book, or my best effort at it. Now that you have reached the end, I want to spend a few pages doing what the chapter structure did not quite allow: talking to you directly, outside the framework, about what I hope you are carrying with you.

You were never broken.

Early in Chapter 1, we introduced the idea that makes children's sleep different from almost any other parenting challenge: the Two-Brain Problem. Your child's sleep is not a standalone system that you observe from the outside. It is a system that is biologically coupled to yours. Your nervous system and your child's nervous system are in active conversation with each other, especially in the early years. The stress hormones in your body travel through the room and register in your child's. The steadiness in your voice at midnight, or the fraying at its edges, is not invisible to a child whose nervous system is still learning what safe feels like. This is not a design flaw. It is attachment biology, and it is one of the things that makes human development remarkable.

But it also means that the standard advice to relax and let your child settle is not as simple as it sounds. Your nervous system is in the room too. Your history with sleep, your anxiety about the next day, your guilt about whether you handled the last waking the right way, your fear that something is fundamentally wrong with your child or with you, all of that is present. The Two-Brain Problem is not just about cortisol and circadian rhythms. It is about the reality that you are a person who is tired, in need of repair, and doing something extraordinarily difficult with very little preparation and very little external support.

When parents come to me with the sense that they have broken their child's sleep, or that they themselves are somehow constitutionally bad at this, I want to tell them what I am telling you now: the biology was always the explanation. The infant who could not settle was not being willful. The toddler who climbed out of the crib every night was not testing you, at least not in the adversarial sense. The school-age child who called out from the dark was not trying to manipulate you. These were nervous systems doing what developing nervous systems do: seeking regulation, seeking proximity, and, with time, support, and a whole lot of repetition, slowly learning to tolerate more independence at night.

You were not broken. You were coupled to a child whose sleep system was still wiring up. The difference between those two things matters enormously, not because one is more forgiving, but because one is true.

There is no perfect parent, and no perfect sleeper

Research on children's sleep is genuinely useful. I believe that, or I would not have spent years reading it and trying to translate it into something practical. But the research is also limited in a way that popular parenting culture does not always acknowledge, and I want to be honest about that limitation before we close.

The studies that shape most sleep guidance were conducted on relatively narrow populations. They tend to oversample two-parent, middle-class households with stable schedules and cooperative infants. They tend to underrepresent the range of real family structures: the single parent running bedtime alone every night, the family where shift work means a different parent handles bedtime on alternating weeks, the household where one child's needs consistently outweigh the other's in the same hour. The research gives us principles. It does not give us a script that works for every household.

More than that: children are not averages. You will read that most infants are capable of longer sleep stretches by a certain age, and your infant will not be most infants. You will read that a particular method shows results within two weeks, and your child will be the one where it takes five. You will read that school-age children generally benefit from a consistent bedtime within a thirty-minute window, and your school-age child will have a week of travel or illness or emotional upheaval that makes the thirty-minute window feel like a theoretical luxury. The gap between the research finding and the specific child in your specific household is not a sign that the research is useless or that your child is a problem. It is just the gap between population-level patterns and individual lives.

There is no perfect sleeper. Not the child of the pediatric sleep specialist, not the infant of the pediatrician who wrote the bestselling sleep book. Sleep is developmental, which means it is in motion. It regresses under stress, illness, travel, and change. It shifts with every developmental transition, and children do not space those transitions to accommodate the routines you have just finished building. The toddler who slept beautifully for three months will stop sleeping beautifully when the next cognitive leap arrives, when a new sibling upends the household, when a caregiver changes, when something at daycare is bothering them in a way they do not yet have words for.

The goal was never perfect sleep. The goal was a family that understands the system well enough to recover more quickly from the inevitable disruptions, and a parent who no longer interprets every setback as evidence of fundamental failure.

You have permission

I want to be explicit about something I believe often goes unsaid in books like this so that I will say it plainly.

You have permission to change your mind.

If you started with a gentle method and it is not working after a reasonable, honest effort, you have permission to try something firmer. The research does not support the idea that a brief period of graduated extinction causes lasting harm to a child who is otherwise securely attached, well-resourced, and not facing other significant stressors. You do not have to choose between your values and your sleep. Sometimes the method you thought you would never use is the one that actually works for your specific child, and adapting to that reality is not a betrayal of anything. It is good parenting.

You also have permission to go the other direction. If you started with extinction because someone you trust told you it was the evidence-based approach, and it is not sitting right with you, you are allowed to stop. You are allowed to decide that your child's distress level is more than you can tolerate, that the responsiveness matters more to you than the timeline, that you are willing to take a slower path if it is a calmer one. That is also good parenting.

You have permission to bedshare if you have weighed the risks, followed the safety guidelines that reduce those risks, and made a deliberate choice that this arrangement is right for your family. The majority of parents worldwide practice bedsharing and have practiced it throughout human history. The American guidelines on infant safe sleep are important, particularly in the first six months and particularly for families with specific risk factors, and this book has covered those carefully. But a family that bedshares thoughtfully, with an older infant or toddler, on a safe surface, without the risk factors the research flags, is not doing something shameful. They are making a choice that works for them.

You have permission to put yourself first sometimes. Parent sleep is not a footnote. It is a variable in the system, and it affects everything else in the system: your emotional regulation during hard nights, your ability to stay consistent with whatever approach you have chosen, your patience in the moments when patience is the whole job. Taking a night when your partner handles it, asking a family member for a stretch of uninterrupted hours, or making the pragmatic choice that gets everyone back to sleep fastest rather than the theoretically optimal choice is not weakness. It is maintenance. A system that runs on no fuel does not run.

You have permission to say that you are tired. Not just physically tired, although that is real and significant, but tired in the way that shows up when you have been giving a great deal and receiving very little back, when the emotional labor of managing another person's sleep while your own sleep suffers has been going on for months or years, when you love your child completely and are also completely exhausted by the particular needs of this season of their life. Both things are true. You are allowed to hold both things at once without deciding that one cancels out the other.

What comes next in this series

This volume covers children's sleep from newborn through eighteen. But the Sleep Engineer series does not stop here.

Volume 3 turns to the family system: how two caregivers build a shared sleep strategy when they approach the problem from different philosophies; how single parents build sustainable routines without a partner to rotate with; how blended families navigate sleep rules across two households with different expectations. Sleep does not happen in a vacuum. It happens inside relationships, and relationships have their own sleep-relevant dynamics that Volume 2 only had room to touch on. Volume 3 goes deeper.

Volume 4 addresses the clinical intersection: what happens when sleep problems are not resolved by routine and method, when there is an underlying diagnosis worth considering, when anxiety, ADHD, or a sensory processing difference is doing most of the work that a sleep strategy cannot fix on its own. That volume is written in close consultation with clinicians who work at that intersection every day, and it is designed to help parents know when to keep working with the tools in books like this one and when to bring in a specialist.

Future volumes in the series will continue to expand the framework outward, following the evidence where it goes and the needs of tired families where they are.

You do not need to read them all. This book is complete on its own. But if the questions it raised led you to other questions, the series will continue to follow those threads.

For the parent reading this at 11 pm, one eye open

You stayed with this to the end. That means something.

The parent who makes it to the final page of a sleep book, in whatever fragments of time they can find between night wakings, early mornings, and the relentless forward motion of caring for a child, is not a failing parent. They are a parent who is still looking, still trying to understand, still putting in effort when the easier thing would be to give up or decide that this is just how it is and stop asking why.

You may not have slept well last night, and you may not sleep perfectly tonight, and the child down the hall may wake before you would like. None of that changes what you have built by paying attention to this: a clearer map of what is happening inside your child's developing brain, a wider menu of tools to draw from, and a little more permission to trust yourself in the dark. The questions you are carrying are not signs that you are lost. They are signs that you are paying attention to something that matters, and that kind of attention, extended over months and years toward a child who needs it, is one of the deepest things a person can do.

Sleep will come for your child, and for you.

You are doing better than you think.

14 Glossary

This glossary collects the key words and concepts you will encounter throughout Children's Sleep. You do not need to memorize any of these terms. Think of this section the way you might think of a neighborhood map: it is here when you want to understand why a certain topic matters, and it is perfectly fine to flip back to it mid-chapter whenever something feels unfamiliar. Each definition is written in plain language because you are a tired parent, not a sleep-lab technician, and plain language is what actually helps at 2 a.m.

AAP (American Academy of Pediatrics): The AAP is the largest professional organization of pediatricians in the United States, and its guidelines shape most doctors' recommendations for infant and child care. On the topic of sleep, the AAP regularly updates its safe-sleep guidance, covering everything from sleep surfaces to room-sharing. When this book references AAP recommendations, it means those guidelines represent a broad scientific and medical consensus on keeping babies safe.

Active Sleep: Active sleep is the earliest term scientists used for what we now often call REM sleep in infants, because babies in this state wiggle, grimace, flutter their eyelids, and make small sounds. It can look alarming to a new parent who wonders whether the baby is waking up, but those movements are a normal part of how the developing brain processes information. Newborns spend roughly half of their total sleep time in active sleep, which is far more than adults do.

Adenosine is a chemical that builds up in the brain the longer you are awake, and that accumulating pressure is one of the main reasons your child grows sleepier as the day goes on. When your child finally falls asleep, adenosine clears, and the pressure resets. Caffeine works by blocking adenosine receptors, which is one reason keeping older children away from caffeinated drinks in the afternoon makes such a difference.

Adenotonsillectomy: An adenotonsillectomy is a surgical procedure that removes the adenoids, the tonsils, or both. In children, enlarged tonsils and adenoids are one of the most common reasons for obstructive sleep apnea, and surgery is often the first-line treatment. Many families report a dramatic improvement in their child's sleep quality and daytime behavior within a few weeks of the procedure.

Attachment: In developmental psychology, attachment refers to the deep emotional bond that forms between a child and their primary caregivers, providing a secure base from which the child can explore the world. Attachment theory, developed by John Bowlby and later by Mary Ainsworth, is relevant to sleep because a securely attached child often finds it easier to self-soothe over time, once trust has been established. Good sleep support and a warm, responsive caregiving relationship reinforce each other rather than conflict.

Bedsharing: Bedsharing means that a parent or caregiver sleeps in the same bed as the child. It is practiced widely around the world and can be driven by cultural tradition, convenience, or a child's needs. This book discusses bedsharing honestly, including the AAP's safety considerations for infants and the reasons many families choose it regardless.

Blue Light: Blue light is the short-wavelength light emitted by sunlight as well as by phone, tablet, and television screens. The brain's internal clock is especially sensitive to blue-wavelength light, and evening exposure signals that it is still daytime, delaying the release of melatonin. Reducing screen use in the hour or two before bedtime is one of the simplest, most evidence-supported steps families can take to make falling asleep easier.

Cataplexy: a sudden, brief loss of muscle tone triggered by strong emotions such as laughter, surprise, or excitement. It is closely associated with narcolepsy and can range from a slight weakness in the face or knees to a complete fall. When it occurs in children, it is sometimes mistaken for clumsiness or fainting, so a proper evaluation is important.

Catnap: A catnap is a very short sleep of roughly 20 to 45 minutes, most commonly seen in newborns and young infants whose sleep cycles are brief. Many parents find the catnap frustrating because the baby seems to wake before feeling truly rested, but short naps are entirely normal in early infancy. As a child's sleep cycles lengthen and consolidate, naps typically extend on their own.

CBT-I (Cognitive Behavioral Therapy for Insomnia): CBT-I is a structured, talk-therapy-based approach to insomnia that helps people identify the thoughts and behaviors keeping them awake, and then replace those patterns with healthier ones. It was developed for adults, but adapted versions are now used for children and teenagers, particularly when anxiety is driving the sleep difficulty. CBT-I for kids might include relaxation exercises, worry journaling before bed, and gradual changes to sleep schedules.

Chair Method: The chair method is a gradual sleep-training approach in which a parent sits in a chair near the child's sleep space and offers quiet reassurance without picking the child up. Every few nights, the parent moves the chair a little farther toward the door, slowly increasing the distance. Many families who want a middle path between full presence and full extinction find the chair method a manageable starting point.

Circadian Rhythm: The circadian rhythm is the body's roughly 24-hour internal clock that regulates when we feel awake or sleepy, when core body temperature rises and falls, and when certain hormones are released. It is set primarily by light and darkness, which is why consistent morning light and dim evening light matter so much. Newborns lack a fully developed circadian clock, which helps explain the exhausting unpredictability of those early weeks.

Confusional Arousal: A confusional arousal is an episode in which a child partially wakes, appears dazed or confused, may cry or thrash, and has no memory of it the next morning. It most often occurs in the first few hours of sleep, when the transition from deep slow-wave sleep is incomplete. These episodes can look dramatic but are generally harmless; the best response is to keep the child safe and wait quietly rather than try to wake them fully.

Cortisol: Cortisol is often called the stress hormone, but it also plays an important role in the healthy sleep-wake cycle by peaking in the early morning to help the body and brain shift into alertness. In the context of sleep challenges, cortisol can become a problem if a child becomes overtired: a late-evening wave of cortisol makes falling asleep harder, not easier, which is the biological explanation for the overtired-but-wired phenomenon many parents know well.

Default Mode Network: The default mode network is a set of brain regions that become particularly active during rest, daydreaming, and the early phases of sleep, when the mind is not focused on a specific task. In children, this network is still maturing, which influences the nature of their dreams and their ability to wind down mentally at bedtime. Understanding that a child's brain does not simply switch off when the lights go out can help parents build more realistic and compassionate bedtime expectations.

Drowsy But Awake: Drowsy but awake is a phrase you will hear often in sleep guidance, and it describes the sweet spot for putting a baby or young child down when they are clearly sleepy but have not yet fully fallen asleep. The goal is to give the child the experience of falling asleep in the place where they will spend the night, so that when they naturally wake between sleep cycles they recognize their surroundings and can drift back off more easily. It sounds simple, but timing it well takes practice and you will not always get it right, which is completely fine.

Enuresis (Bedwetting): Enuresis is the medical term for bedwetting, which means urinating during sleep at an age when dryness might reasonably be expected. Nighttime bladder control develops at very different rates in different children, and bedwetting is considered a normal variant up to age five or even later. Most children outgrow it without any intervention, though several treatment options exist for older children for whom it causes distress.

Extinction, in the context of sleep training, refers to allowing a child to fall asleep without parental intervention, providing no external response to their crying so that the behavior gradually decreases on its own. The term comes from behavioral psychology and is often informally referred to as "cry-it-out". Research consistently shows it is effective and does not cause lasting harm to attachment or emotional development, though it is not the right fit for every family's temperament or circumstances.

Ferber Method: The Ferber method, named after pediatric sleep specialist Dr. Richard Ferber, is a graduated form of extinction in which parents check on their child at increasing time intervals rather than staying entirely away. A parent might check in after 3 minutes, then 5, then 10, offering brief verbal reassurance without picking the child up. This approach is sometimes called graduated extinction or check-and-console, and many parents find the periodic check-ins easier to sustain emotionally than full extinction.

Ferritin: Ferritin is a protein that stores iron in the body, and blood ferritin levels are the most common way doctors assess iron status. Low ferritin, even in the absence of full anemia, has been linked to restless legs syndrome and periodic limb movement disorder in children, both of which can severely fragment sleep. If your child complains of uncomfortable sensations in their legs at night or seems unable to stay still in bed, asking your pediatrician to check ferritin levels is a reasonable first step.

Four-Month Progression: The four-month progression, sometimes called the four-month sleep regression, is a permanent neurological shift in how a baby's sleep is organized. Around this age, infant sleep begins to cycle through lighter and deeper stages in a pattern more similar to adult sleep, and babies who were previously sleeping in longer stretches may suddenly wake more frequently. Calling it a progression rather than a regression is more accurate because the brain is actually maturing, not going backward, even though the sleepless nights feel very much like things have gotten worse.

Homeostatic Sleep Pressure: Homeostatic sleep pressure is the biological drive to sleep that builds the longer a person is awake, driven largely by the accumulation of adenosine in the brain. It is one of two main systems that govern when we sleep, working alongside the circadian clock. The interplay between these two systems determines the ideal bedtime and wake window for your child at any given stage of development.

Hypnogram: a chart that maps a person's sleep stages across the entire night, showing when they moved in and out of light sleep, deep sleep, and REM sleep. Sleep specialists use hypnograms to identify patterns like too little deep sleep, too many awakenings, or abnormal movements during the night. You do not need to produce one at home, but understanding what a healthy childhood hypnogram looks like can help you make sense of the sleep stage information throughout this book.

Melatonin is a hormone produced by the pineal gland in the brain in response to darkness, serving as a biological signal that nighttime has arrived. It does not force sleep but rather shifts the body toward sleepiness. As a supplement, melatonin is widely used to adjust sleep timing, particularly for travel or circadian phase issues. However, its long-term use in children is an evolving area of research and worth discussing with your pediatrician before starting.

Narcolepsy: Narcolepsy is a neurological disorder in which the brain cannot reliably regulate the boundary between wakefulness and sleep, causing excessive daytime sleepiness and, in some cases, sudden episodes of muscle weakness triggered by emotion. It results from the loss of neurons that produce the wake-promoting chemical hypocretin, also known as orexin. Narcolepsy is rare in children but can go undiagnosed for years because its symptoms are sometimes attributed to laziness or behavior problems.

Night Terror: A night terror, also called a sleep terror, is a partial arousal from deep slow-wave sleep during which a child may scream, cry, sit up, or appear terrified but is not awake and will not remember the episode. Night terrors are most common between ages 3 and 8 and are more common in children who are overtired, ill, or dealing with a disrupted sleep schedule. Unlike a nightmare, a night terror happens in the first third of the night, and the child is unresponsive to comfort during the episode.

Nightmare: a disturbing dream that occurs during REM sleep and that the child can usually remember and describe upon waking. Nightmares are a normal part of childhood development, especially between ages 3 and 6 when imagination is blossoming, and fears are becoming more complex. They are different from night terrors in that the child is actually awake afterward, is aware of your presence, and genuinely benefits from your comfort.

Obstructive Sleep Apnea (OSA): OSA is a condition in which the airway partially or fully collapses during sleep, causing pauses in breathing, oxygen dips, and frequent brief arousals. In children, the most common cause is enlarged tonsils and adenoids, though obesity and certain anatomical features also play a role. Untreated OSA in children is associated with behavioral problems, difficulty concentrating, and impaired growth, making early identification and treatment important.

Oxytocin: Oxytocin is a hormone and neurotransmitter sometimes called the bonding hormone, released during physical closeness, breastfeeding, and warm social interaction. It promotes feelings of calm and safety in both parent and child, which is one reason that rocking, nursing, or a long cuddle before bed can be such an effective part of a soothing bedtime routine. The relationship between oxytocin and sleep is a reminder that the warm rituals of bedtime are not just emotionally meaningful; they are physiologically useful.

Pacifier: A pacifier is a nipple-shaped device given to a baby for non-nutritive sucking, and it has been studied extensively in the context of infant sleep. The AAP recommends offering a pacifier at naptime and bedtime for infants up to 12 months because consistent use has been associated with a reduced risk of sudden infant death syndrome. One consideration many parents face is the "pacifier association," where a baby wakes at night needing the pacifier replaced because sucking was the only way they learned to fall asleep.

Pantley (No-Cry) Method: The Pantley method, developed by author Elizabeth Pantley in her No-Cry Sleep Solution, is a gradual approach that aims to shift a baby's sleep associations without prolonged crying. The core technique involves offering your usual comfort (nursing, rocking, pacifier) and then slowly removing it just before the baby is fully asleep, repeating over many nights until the baby can complete the transition to sleep independently. Progress is slower than with extinction-based methods, but many parents strongly prefer the approach because it aligns with their instinct to respond to their child.

Parasomnia: a broad term for any unusual or disruptive behavior that occurs during sleep or during transitions between sleep and wakefulness. Night terrors, sleepwalking, confusional arousals, and sleep-related eating disorder all fall under this umbrella. Most childhood parasomnias are benign, tend to run in families, and diminish as the brain matures.

Phase Delay (Adolescent): Phase delay refers to the natural biological shift during puberty in which the circadian clock shifts later, making teenagers feel genuinely alert later in the evening and unable to fall asleep early, no matter how hard they try. This is not laziness or a bad habit; it is a documented neurological change driven by hormonal shifts during adolescence. Early high school start times that conflict with this delayed biology are associated with sleep deprivation, reduced academic performance, and increased risk of depression in teens.

Polysomnography: Polysomnography is the gold-standard sleep study conducted in a clinical setting, where a person sleeps overnight. At the same time, sensors track brain activity, eye movements, breathing, oxygen levels, heart rate, and limb movements simultaneously. Pediatric sleep specialists use it to diagnose conditions like sleep apnea, narcolepsy, and certain parasomnias. It sounds daunting, but most children, especially when a parent stays the night with them, adjust quickly to the sensors and fall asleep reasonably well.

Process C refers to the circadian component of sleep regulation. This roughly 24-hour biological clock promotes wakefulness during the day and sleepiness at night, driven by light exposure and internal timing. It is the second of the two main forces governing when we sleep, working alongside Process S. In infants, Process C takes several weeks to months to become fully functional, which is why newborn sleep feels so random and why deliberately managing light and darkness becomes more useful around 8 to 12 weeks.

Process S refers to the sleep-homeostatic component of sleep regulation, the pressure to sleep that builds throughout waking hours and dissipates during sleep. The S stands for sleep pressure, and it is driven by the gradual accumulation of adenosine in the brain. Understanding Process S helps explain why the timing and length of naps matter so much: a nap that is too late or too long relieves enough sleep pressure that bedtime becomes a battle.

Quiet Sleep: Quiet sleep is the term used for the non-REM, deep, restful sleep seen in infants before their sleep architecture matures into the more adult-like stages. During quiet sleep, a baby is very still, breathing is regular, and they are difficult to rouse. It corresponds roughly to the slow-wave, or deep, sleep seen in older children and adults, and is the state associated with the greatest physical rest and growth.

REM Sleep: REM stands for rapid eye movement, and REM sleep is the sleep stage most associated with vivid dreaming, emotional memory processing, and creative thinking. During REM, the body is essentially paralyzed to prevent acting out dreams, even though the brain is highly active. Children spend a larger proportion of their sleep in REM than adults do, reflecting how much heavy neural building and pruning happens in early childhood.

Restless Legs Syndrome (RLS) is a neurological condition characterized by uncomfortable sensations in the legs, most often described as a creeping, tingling, or irresistible urge to move, that worsen at rest and typically peak in the evening. In children, it can make settling at bedtime extremely difficult and is often misidentified as a behavior problem. Low iron stores are one of the most modifiable risk factors, and iron supplementation is sometimes the first treatment tried.

Room-Sharing: The baby sleeps in the same room as the parent, but on a separate sleep surface, such as a crib, bassinet, or bedside sleeper. The AAP recommends room-sharing without bedsharing for at least the first six months of an infant's life and ideally for the first year. Beyond safety, many families find room-sharing to be a practical convenience, making nighttime feeding and monitoring easier during those early months.

Safe Sleep: Safe sleep refers to the full set of evidence-based practices designed to reduce the risk of sudden unexpected infant death during sleep, particularly in the first year of life. The core recommendations include placing babies on their backs, on a firm flat surface, with no soft bedding or objects in the sleep space, and in the parents' room but not in the parents' bed. These are not suggestions made lightly; they reflect decades of research and have contributed to a significant decline in infant sleep-related deaths.

Sleep Architecture: Sleep architecture refers to the overall structure of sleep across a full night, including how much time is spent in each stage and the sequence in which those stages occur. A healthy night of sleep moves through multiple cycles of lighter sleep, deep slow-wave sleep, and REM sleep, with the proportion of deep sleep heavier in the early part of the night and REM heavier in the later part. Sleep architecture changes dramatically from birth through adolescence, which is why age-appropriate expectations matter so much in children's sleep.

Sleep Cycle: A sleep cycle is one complete pass through the major stages of sleep, from light sleep through deep sleep and back up through REM. In adults, a cycle lasts roughly 90 minutes; in newborns, it is closer to 45 to 50 minutes, which explains why babies so often stir at predictable intervals. Children naturally rouse slightly between cycles, and the ability to reconnect with sleep during that transition is at the heart of what sleep training is designed to teach.

Sleep Onset Latency: Sleep onset latency is the amount of time it takes a person to fall asleep from the moment they lie down with the lights off and intend to sleep. A typical healthy range for children is roughly 15 to 30 minutes. Very short latency (falling asleep in under five minutes) can actually signal excessive sleep deprivation, while very long latency often points to circadian misalignment, anxiety, or an overly stimulating pre-bed environment.

Sleep Training: Sleep training is an umbrella term for any deliberate approach a parent takes to help their child develop the ability to fall asleep independently and to resettle between sleep cycles during the night. It covers everything from very gradual, no-cry methods to quicker extinction-based approaches. There is no single right method; the most important factor is choosing an approach that is safe, age-appropriate, and easy to follow consistently.

Social Jet Lag: Social jet lag describes the misalignment between a person's biological sleep timing and the social schedule imposed by school, work, or activities. A teenager whose body clock wants them to be asleep at midnight but who must wake at 6 a.m. on school days and then sleep until 10 a.m. on weekends is experiencing social jet lag. The resulting chronic partial sleep deprivation carries real costs to mood, metabolism, and learning.

Suprachiasmatic Nucleus (SCN): The suprachiasmatic nucleus is a tiny cluster of nerve cells in the hypothalamus at the base of the brain that functions as the master clock of the circadian system. It receives light signals directly from the retina and uses them to synchronize biological rhythms throughout the body. When sleep scientists talk about "resetting the clock" with morning light or melatonin, the SCN is the structure they are targeting.

Swaddling: Swaddling is the practice of wrapping a young baby snugly in a blanket to help them feel secure, reduce the startle reflex that can disrupt sleep, and sometimes extend sleep. It is most appropriate for the first two to three months of life, and parents are advised to stop swaddling when their baby shows any signs of rolling over because a swaddled baby who rolls face-down cannot push up safely. A proper swaddle is snug around the arms but leaves room for the hips to move.

Transitional Object: A transitional object is a comfort item, most often a soft toy, blanket, or stuffed animal, to which a child develops a strong emotional attachment and that helps them manage separation from a caregiver at bedtime. The concept was introduced by pediatrician and psychoanalyst D.W. Winnicott, who called it a "security object." Most children develop an attachment to a transitional object naturally between 8 and 18 months, and there is no need to rush or discourage it.

Wake Window: A wake window is the amount of time a child can comfortably stay awake between sleep periods, whether between naps or between the last nap and bedtime. Wake windows change substantially over the first few years of life, and matching your timing to a developmentally appropriate window is one of the most practical tools for preventing an overtired or under-tired child at bedtime. A child put down before their wake window is up may not be sleepy enough; one put down significantly after it may be overtired and paradoxically harder to settle.

15 The Twelve-Night Family Sleep Reset

This appendix is the whole book on one page: one night, one chapter, one small experiment. You do not have to follow the order. You do not have to do all twelve. Pick the night that fits your child's age and your tiredness, and try just that one. If it helps, do it again tomorrow; if it does not, move on without guilt. The point is not perfection. The point is to feel, in your own house, that small changes in how sleep is engineered add up to better nights.

Night 1: The Two-Brain Problem

For this age range: All ages

Sync one moment of calm with your child tonight, even if it is sixty seconds of slow breathing together at bedtime. Your nervous system is the cue their nervous system is reading.

Night 2: The Newborn Brain

For this age range: 0 to 3 months

Spend the next twenty-four hours noticing your newborn's sleep cycles instead of fighting them. Feed, soothe, repeat. Stop trying to make tonight sleep like a four-month-old's.

Night 3: The Clock Starts Ticking

For this age range: 3 to 6 months

Pick one anchor for the body clock and run it for the next three days at the same time: morning sunlight, a consistent first feed, or a dim, predictable bedtime cue.

Night 4: Naps, Schedules, and Routines

For this age range: 4 to 18 months

Stop watching the clock and start watching the wake window. Track yawns, eye-rubs, and stares, then offer a nap before the second sleepy cue.

Night 5: The Toddler Bedtime Battle

For this age range: 1 to 3 years

Pre-decide how many books, how many sips, how many hugs. Tell your toddler the plan before bedtime begins, then hold the limit with warmth rather than anger.

Night 6: Preschool Sleep

For this age range: 3 to 5 years

If your child has dropped the nap and falls apart by dinner, add a thirty-minute quiet time after lunch tomorrow. Books, lights low, no screens, no demand for sleep.

Night 7: School-Aged Sleep

For this age range: 6 to 11 years

Move bedtime fifteen minutes earlier for the next three nights. Cut screens thirty minutes earlier than that. Watch what happens to the morning.

Night 8: The Teenage Brain

For this age range: 12 to 18 years

Negotiate, do not dictate. Ask your teen what they would change about their own sleep, and offer to handle one piece of friction so they can try.

Night 9: Routines That Hold

For this age range: All ages

Build the smallest possible bedtime routine that can survive a bad day: three steps, ten minutes, the same order every single night, even when you are exhausted.

Night 10: Anxiety, Worries, and the Bedtime Mind

For this age range: 4 years and up

Create a worry time outside of bedtime. Five minutes, paper and pencil, before dinner. Anything that surfaces at night gets parked there for tomorrow.

Night 11: Troubleshooting and Red Flags

For this age range: All ages

Spend one night this week as a careful observer. Notice snoring, mouth-breathing, restless legs, sleepwalking, or anything that makes you uneasy. If something stands out, call your pediatrician this week.

Night 12: Sleep Myths Decoded

For this age range: All ages

Pick one sleep belief you have inherited from family, social media, or a previous generation. Hold it next to what you have read in this book. Decide tonight if it still serves your family.

Twelve nights is not a deadline. It is a permission slip. Start where your family is. Borrow whichever night feels useful. Skip the rest. Come back when you need to. The science is here when you want it. The mercy is yours to give.

www.ingramcontent.com/pod-product-compliance
Lightning Source LLC
LaVergne TN
LVHW052033200926
848481LV00005B/1661